Shallow Subduction Zones: Seismicity, Mechanics and Seismic Potential
Part I

Edited by
Renata Dmowska
Göran Ekström

1993

Birkhäuser Verlag
Basel · Boston · Berlin

Reprint from Pure and Applied Geophysics
(PAGEOPH), Volume 140 (1993), No. 2

Editors' addresses:

Renata Dmowska
Division of Applied Sciences
Harvard University
Cambridge, MA 02138
U.S.A.

Göran Ekström
Department of Earth and Planetary Sciences
Harvard University
Cambridge, MA 02138
U.S.A.

A CIP catalogue record for this book is available from the Library of Congress,
Washington D.C., USA

Deutsche Bibliothek Cataloging-in-Publication Data

Shallow subduction zones: seismicity, mechanics and seismic
potential / ed. by Renata Dmowska and Göran Ekström. –
Basel ; Boston ; Berlin : Birkhäuser.
NE: Dmowska, Renata [Hrsg.]
Pt. 1. Fortdruck aus Pure and applied geophysics (PAGEOPH),
    Vol. 140, 1993. – 1993
    ISBN 3-7643-2962-9 (Basel . . .)
    ISBN 0-8176-2962-9 (Boston)

© 1993 Birkhäuser Verlag, P.O. Box 133, CH-4010 Basel, Switzerland
Printed on acid-free paper produced from chlorine-free pulp
Printed in Germany
ISBN 3-7643-2962-9
ISBN 0-8176-2962-9

9 8 7 6 5 4 3 2 1

Contents

PAGEOPH, Vol. 140, No. 2 (1993)

0033–4553/93/020179–03$1.50 + 0.20/0
© 1993 Birkhäuser Verlag, Basel

Introduction

RENATA DMOWSKA[1,2] and GÖRAN EKSTRÖM[2]

The generation of large earthquakes in subduction zones presents a broad spectrum of challenging scientific problems in the Earth Sciences. From plate tectonics we know that subduction zones are locations where the oceanic lithosphere deforms and sinks into the mantle beneath the neighboring plate, and that they are a primary manifestation of mantle convection and dynamics. At shallow depths, the subduction processes give rise to volcanism and to most of the world's seismicity, including the largest earthquakes. In addition, from a geological perspective, the shallow parts of subduction zones are also the areas of intense orogenesis and crustal deformation.

In many regions subduction produces large amounts of seismicity, reflecting the rapid strains that occur both in the interplate contact zone, and within the colliding plates themselves; in other areas, the subduction process appears to proceed without any significant seismic activity or visible evidence of strain accumulation. This difference in behavior, which can be seen when comparing different subduction zones or different segments within a subduction zone, is controlled by the mechanics of almost rigid plates, a viscous substrate, and of the contact zones between the plates.

Progress in observational as well as theoretical investigations of subduction zones over the last decade or two has increased our fundamental knowledge of tectonic processes in these regions, but has also revealed complex and poorly understood behavior which suggests that no simple laws will soon be found that can satisfactorily explain every aspect of the subduction process. An understanding of the complex physics governing the observed diversity of interplate behavior in subduction zones is the common goal of the studies presented in this special issue, and each provides a different approach which contributes to our understanding and ability to predict the seismicity, or lack thereof, in different subduction zones.

[1] Division of Applied Sciences, Harvard University, Cambridge, MA 02138, U.S.A.
[2] Department of Earth and Planetary Sciences, Harvard University, Cambridge, MA 02138, U.S.A.

In the first volume of this topical issue, YU *et al.* show how large scale relative plate motion is partitioned between the interplate contact and within the overriding plate, and how this is a common phenomenon that can be observed in most subduction zones. However, there are systematic geographical differences, and the study provides some observational constraints on what may govern the coupling between subducting and overriding plates. LIU and MCNALLY apply new modeling techniques to address the question of what the stress level might be within the colliding plates and across the interplate contact. Their focus is on earthquakes near the outer rise, where a correlation exists between the phase of the seismic cycle, and the depth of faulting and type of focal mechanism that is observed. With a given rheological model of the oceanic lithosphere, constraints can be placed on the variations in compressive stresses transmitted in the plate.

In the third paper of this volume KISSLINGER reviews the wealth of observations of seismicity in subduction zones that has been collected by local and regional networks in these areas. Data from these networks have been essential for determining the seismic velocity structure in subduction zones, as well as for obtaining accurate locations of hypocenters. Also, the data have allowed detailed geographical mapping of seismic activity down to the microearthquakes size, and these seismicity patterns have been used for earthquake prediction and the identification of asperities on the fault plane.

The next three papers attempt to evaluate the seismic potential of different subduction segments. ZÚÑIGA *et al.* analyze the aftershock sequence of an $M_s = 6.9$ earthquake in the Acapulco–San Marcos segment of the Mexican subduction zone and discuss the implications of their observations for the seismic potential of that area. PAPADIMITRIOU evaluates the validity of the time-predictable model for strong, shallow earthquakes along the western coast of South and Central America, and then estimates the seismic potential of particular segments within the time window of 1992–2002. COMTE and SUÁREZ analyze the complex spatio-temporal behavior of seismicity of two well recognized seismic gaps in southern Peru and northern Chile in an attempt to infer their current seismic potential. Even if observed complexities and the available database of only ~ 30 years do not allow for a firm estimate of the seismic potential in these areas, this is the most comprehensive study of its kind of these zones and will be invaluable in future studies.

The final two papers in this volume represent case studies. HOUSTON *et al.* analyze the anomalous (unusual focal mechanism, relatively high stress drop and short duration) and thus controversial $M_w = 7.7$ earthquake that occurred on October 20, 1986 in the Kermadec segment, with the epicenter approximately 200 km south of the intersection of the Louisville Ridge and the Tonga-Kermadec trench. To evaluate the tectonic significance of that earthquake the authors perform a comprehensive analysis of source parameters using surface- and body waves and relocated aftershocks. They conclude that this event represents an intraplate faulting within the downgoing plate, and appears to be associated with segmentation of the subducting plate produced by forces related to the subduction of the Louisville Ridge.

To analyze fault plane heterogeneities in the northern Solomon Islands subduction segment and their association with rupture characteristics in general and the existence of earthquake doublets in particular, XU and SCHWARTZ study in detail two sets of doublets, from 1974 and 1975, and then relocate 85 underthrusting events in the area. The authors find that few smaller magnitude events overlap asperity regions, and that the majority of small magnitude underthrusting earthquakes occupy a segment that has never experienced a magnitude greater than 7.0 earthquake in the historic times.

It will be of great value to society when seismologists and geophysicists are able to monitor and predict the pattern of geophysical phenomena associated with subduction; this issue presents a modest step towards this goal.

PAGEOPH, Vol. 140, No. 2 (1993)

0033–4553/93/020183–28$1.50 + 0.20/0
© 1993 Birkhäuser Verlag, Basel

Slip Partitioning along Major Convergent Plate Boundaries

GUANG YU,[1] STEVEN G. WESNOUSKY,[1] and GÖRAN EKSTRÖM[2]

Abstract—Along plate boundaries characterized by oblique convergence, earthquake slip vectors are commonly rotated toward the normal of the trench with respect to predicted plate motion vectors. Consequently, relative plate motion along such convergent margins must be partitioned between displacements along the thrust plate interface and deformation within the forearc and back-arc regions. The deformation behind the trench may take the form of strike-slip motion, back-arc extension, or some combination of both. We observe from our analysis of the Harvard Moment Tensor Catalog that convergent arcs characterized by back-arc spreading, specifically the Marianas and New Hebrides, are characterized by a large degree of slip partitioning. However, the observed rates, directions, and location of back-arc spreading are not sufficient to account for degree of partitioning observed along the respective arcs, implying that the oblique component of subduction is also accommodated in part by shearing of the overriding plate. In the case of the Sumatran arc, where partitioning is accommodated by strike-slip faulting in the overriding plate, the degree of partitioning is similar to that observed along the Marianas, but the result is viewed with caution because it is based on a predicted plate motion vector that is based on locally derived earthquake slip vectors. In the case of the Alaskan-Aleutian arc, where back-arc spreading is also absent, the degree of partitioning is less and rotation of slip vectors toward the trench normal appears to increase linearly as a function of the obliquity of convergence. If partitioning in the Alaskan-Aleutian arc is accommodated by strike-slip faulting within the upper plate, the positive relationship between obliquity of convergence and the rotation of earthquake slip vectors to the trench normal may reflect that either (1) the ratio of the depth extent of strike-slip faults behind the trench Z_s to the subduction thrust Z_t increases westward along the arc, (2) the dip of the subduction thrust increases westward along the arc, or (3) the strength of the subduction thrust decreases westward along the arc.

Key words: Slip partitioning, strain partitioning, plate tectonics, convergent plate boundaries.

Introduction

It has been observed that the azimuths of slip vectors for shallow plate boundary thrust earthquakes commonly do not accurately reflect the predicted global plate motion vectors (FITCH, 1972; BECK, 1983, 1989, 1991; JARRARD, 1986; EKSTRÖM and ENGDAHL, 1989; DeMETS *et al.*, 1990; McCAFFREY, 1990, 1991, 1992). Rather, slip vectors for thrust earthquakes are generally rotated toward the

[1] Center for Neotectonic Studies, University of Nevada, Reno, NV 89557, U.S.A.
[2] Department of Earth and Planetary Sciences, Harvard University, Cambridge, MA 02138, U.S.A.

normal of the trench with respect to predicted plate motion vectors (JARRARD, 1986; EKSTRÖM and ENGDAHL, 1989; BECK, 1989, 1991; DEMETS *et al.*, 1990; MCCAFFREY, 1991, 1992; JONES and WESNOUSKY, 1992). Assuming that relative plate motion models are correct, a consequence of this observation is that relative plate motion must be partitioned between displacement along the thrust plate interface and deformation in the forearc or back-arc region. We refer to this phenomenon as slip partitioning. More specifically, we define the slip partitioning angle ψ as the angle between the azimuth of the earthquake slip vector and the trench normal, and the obliquity θ as the angle between the plate motion direction and the trench normal (Figure 1). Hence, when slip vectors are rotated to the trench normal away from the predicted plate motion ($\psi = 0$), slip partitioning is considered complete. Conversely, when earthquake slip vectors are parallel to the predicted plate motion ($\psi = \theta$), there is no partitioning of plate motion. In this paper, we provide a summary of the degree of slip partitioning observed along the major convergent plate boundaries of the world (Figure 2).

Method and Data

The principal data base for this study is the Harvard Moment Tensor Catalog for the period 1977 to 1990 (for a complete list of references to this Catalog,

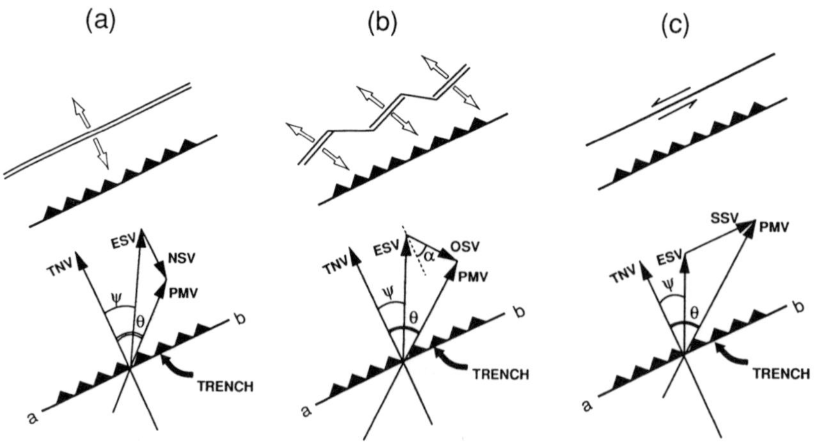

Figure 1

Diagram illustrates the common relationship between predicted plate motion vectors (PMV), the azimuth of shallow thrust earthquake slip vectors (ESV), and orientation of subduction zone, as defined by a trench normal vector (TNV). The line with solid triangles represents a segment of trench, the angle θ defines the obliquity of plate convergence and ψ is the slip partitioning angle between the azimuth of earthquake slip vector (ESV) and the trench normal (TNV). The observed rotation of earthquake slip vectors (ESV) toward the trench normal (TNV) with respect to the predicted plate motion vector (PMV) may be accounted for by (a) back-arc spreading parallel to the trench normal vector (vector NSV), (b) back-arc spreading at an oblique angle α with respect to the trench normal (vector OSV), or (c) strike-slip faulting behind the trench (vector SSV).

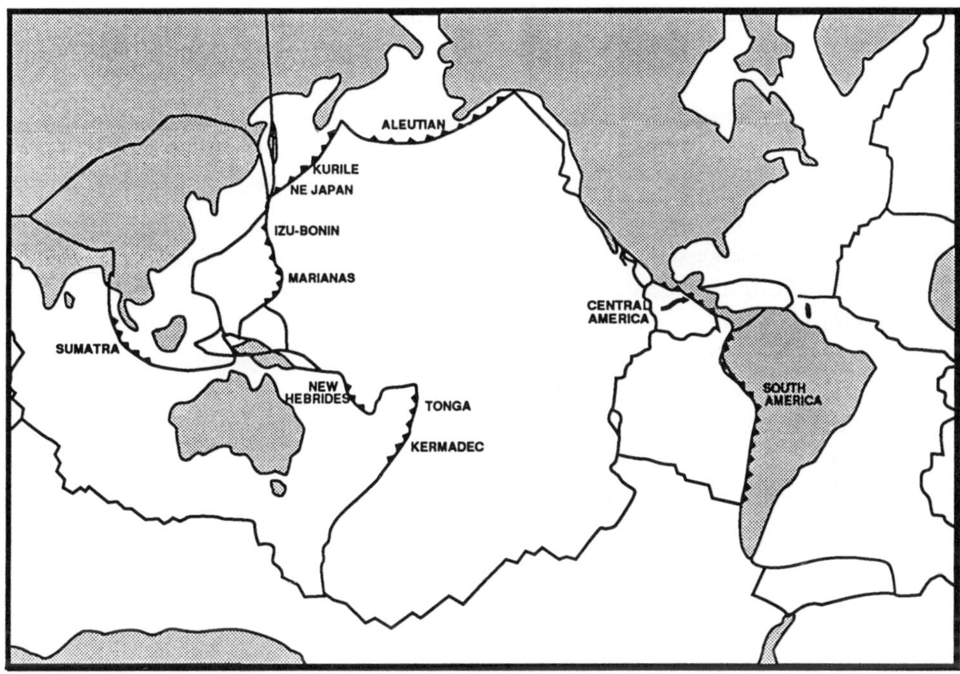

Figure 2

Mercator projection of major plate boundaries. Subduction zones considered in this study are labeled and marked by solid triangles.

see DZIEWONSKI *et al.*, 1991). The Catalog provides earthquake focal mechanisms for most earthquakes of $M_w \geq 5.0$. Along each convergent plate boundary, shallow earthquakes (0–40 km) showing thrust focal mechanisms are selected. Toward determining an estimate of slip partitioning, we compare the azimuth of the earthquake slip vector to the azimuth of the predicted plate motion vector at the same site. The azimuth of the slip vector for each event is obtained by rotating the slip vector about the strike of focal plane into the horizontal. The relative plate motion directions are calculated from the global motion model NUVEL-1 proposed by DEMETS *et al.* (1990), and from SENO *et al.* (1987) for Izu-Bonin and Mariana trenches. The trench normal is inferred from digital bathymetric maps with a 5 minute by 5 minute resolution for each island arc by taking the normal to a curve interpolated between points of maximum depth along each trench at a spatial interval of 15 minutes of latitude or longitude, respectively, depending on the orientation of the trench.

The map view of slip vectors with respect to the predicted relative plate motion vectors along the Aleutian trench in Figure 3a serves to further illustrate the data analysis. The variation of the trench normal vector, the predicted plate motion vector, and the azimuth of individual earthquake slip vectors are plotted in Figure 3b as a function of longitude along the trench. Both Figures 3a and 3b show that slip vectors are generally rotated away from the predicted plate motion direction and fall between the plate motion direction and trench normal. The relationship of the slip partitioning angle ψ for each event and the angle of the obliquity of subduction θ is further shown in Figure 3c. Solid circles in both Figures 3b and 3c represent the events with moment M_0 greater than 1.5×10^{24} dyne-cm, dip angle less than $40°$ and strike within $\pm 20°$ of the strike of the trench. Open circles are the events along the trench which do not satisfy these criteria. The events marked by open circles show greater scatter as a function of θ. The events of lesser seismic moment are generally more likely to be affected by local processes rather than reflect plate motion, as compared to events of greater seismic moment. Similarly, events with steep dip angle ($>40°$) and strike much different ($\pm 20°$) from the strike of the trench are less likely to reflect the general characteristics of plate motion at subduction zones. Source parameters of the smaller events are associated with larger uncertainties because of the low signal to noise ratio in the seismograms. For these reasons we infer that it is the events which satisfy the above three criteria that are likely to reflect best any systematic relationship between the slip partitioning angle ψ and obliquity θ of subduction. It is with these same criteria that we constructed the same sequence of plots for the South American (Figure 4), Central American (Figure 5), Kurile and Japan (Figure 6), Sumatran (Figure 7), Tonga and Kermadec (Figure 8), Izu-Bonin and Marianas (Figure 9), and New Hebrides (Figure 10) island arcs. Thus, the data in Figures 3 to 10 serve as the observational data base for this study.

Observations

Arcs without Back-arc spreading

Among the subduction zones considered, the Alaskan-Aleutian, South American, Central American, Kurile, Japan and Sumatran do not show evidence of back-arc spreading. The obliquity of subduction θ along the Alaskan-Aleutian arc ranges from about $0°$ to $80°$ (Figure 3). Values of slip partitioning angle ψ appear to systematically increase as a function of obliquity angle θ and define a slope of less than 1 (Figure 3c).

Along South America (Figure 4), values of obliquity range to only $40°$ and values of slip partitioning angle tend to fall below the line $\psi = \theta$. There also appears to be a tendency for slip partitioning angle ψ to increase as a function of obliquity θ, but because the azimuth of the NUVEL-1 plate motion vector for this plate boundary is largely constrained from locally derived earthquake slip vectors and the scatter in data is large, the result in equivocal (Figure 4c).

Aleutians
60N
55N
50N
45N
165E
170E
175E
180E
175W
170W
165W
160W
155W
150W

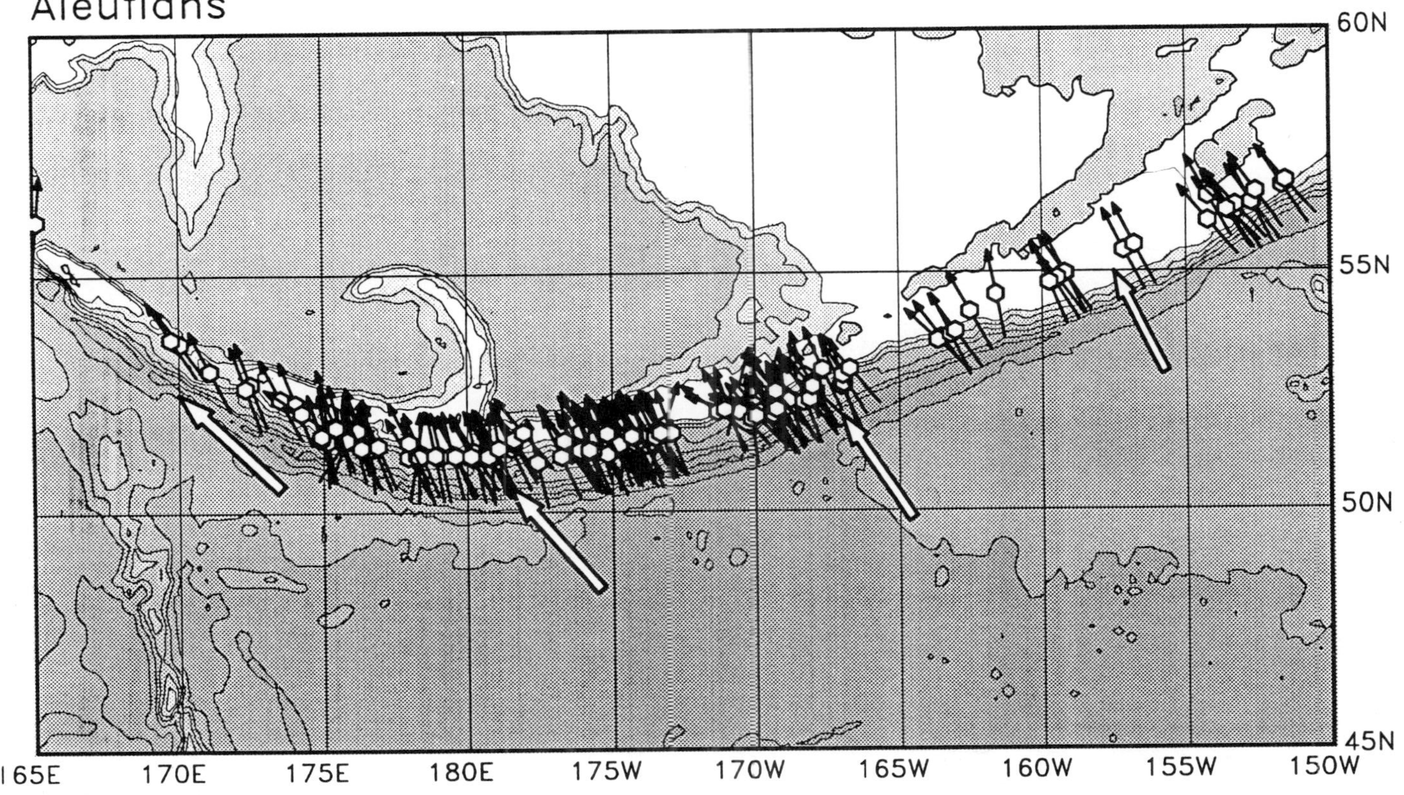

Figure 3(a)

Aleutians

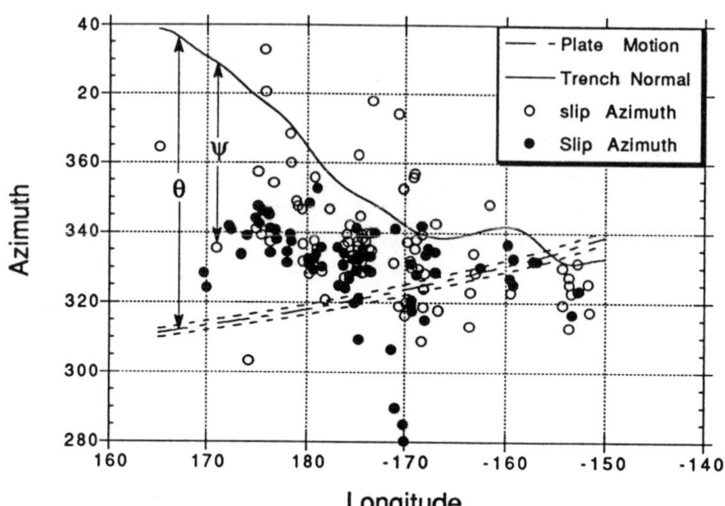

Figure 3(b)

Aleutians

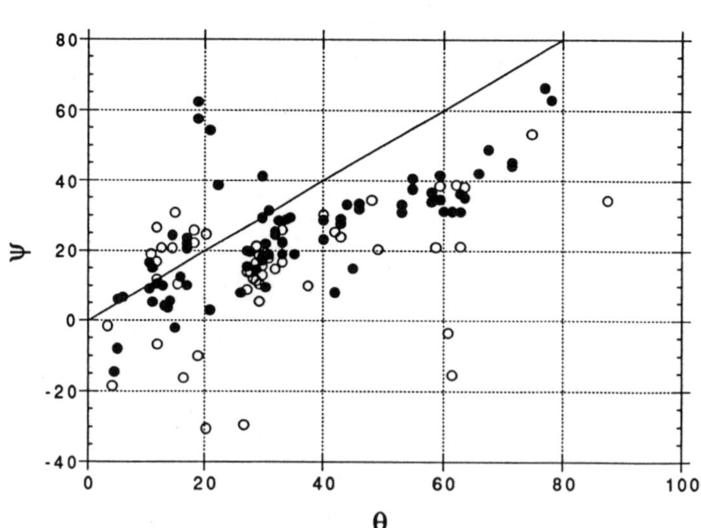

Figure 3(c)

Figure 3

(a) Map view of earthquake slip vectors (solid arrows), which are obtained by rotating the slip vectors about the strike of focal plane into the horizontal plane, for shallow thrust earthquakes (0–40 km) along the Aleutians. Large open arrows indicate the directions of predicted relative plate motion calculated from NUVEL-1 model. (b) A plot of the azimuths of earthquake slip vectors, plate motion vector and the trench normal vector versus longitude along Aleutians. Solid line marks the azimuth of the trench normal. Predicted plate motion directions and the uncertainties are represented by dot-dashed line

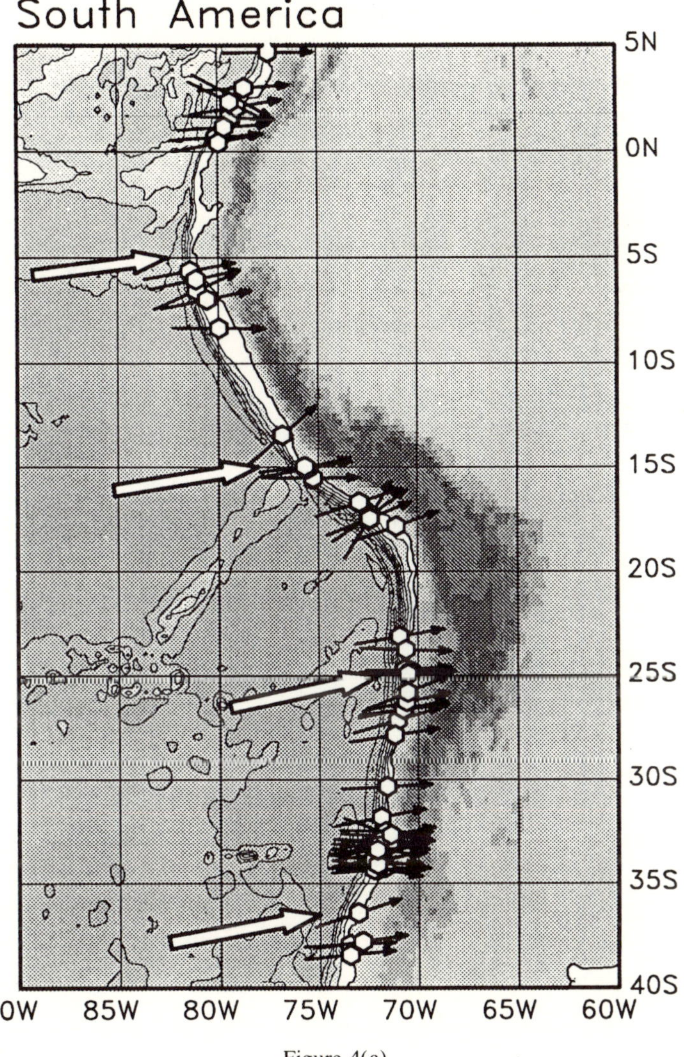

Figure 4(a)

and dashed lines, respectively. Slip vector azimuths are marked by solid circles for events with $M_0 \geq 1.5 \times 10^{24}$, dip angle less than $40°$, and for which the difference between the strike of focal plane and the strike of trench is less than $\pm 20°$. Open circles are events along the trench which do not satisfy the three criteria. Obliquity θ is defined as the angle between the trench normal and predicted plate motion vectors. The partitioning angle ψ is defined by the angle between the azimuth of the slip vector and the trench normal. (c) A plot of ψ versus θ for Aleutians. The line $\psi = \theta$ (slope of 1) is also plotted for reference.

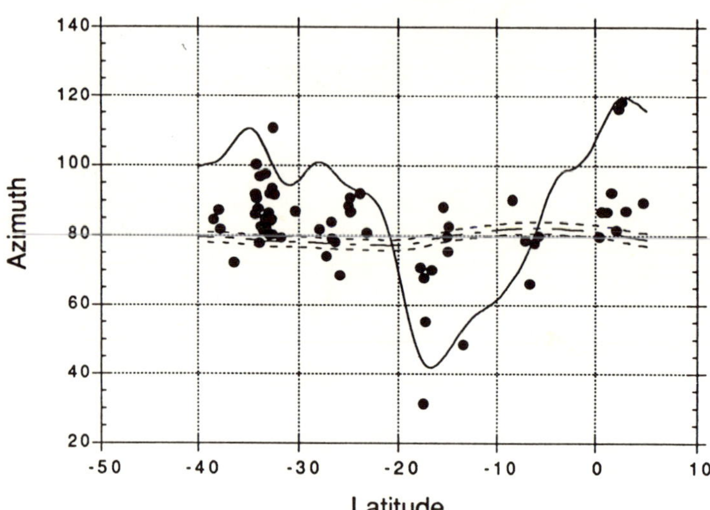

Figure 4(b)

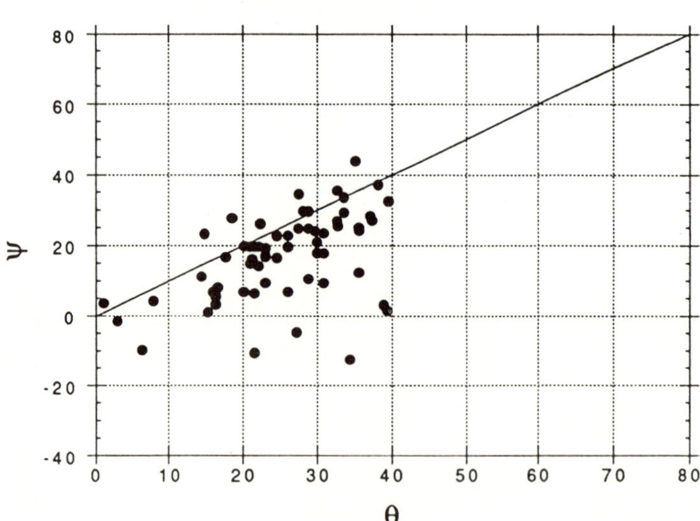

Figure 4(c)

Figure 4

Observations for South America. See Figure 3 caption for explanation.

20N
15N
10N
5N
85W
90W
95W
100W
Central America
COCOS RIDGE
TEHUANTEPEC RIDGE
Caribbean-North America Boundary

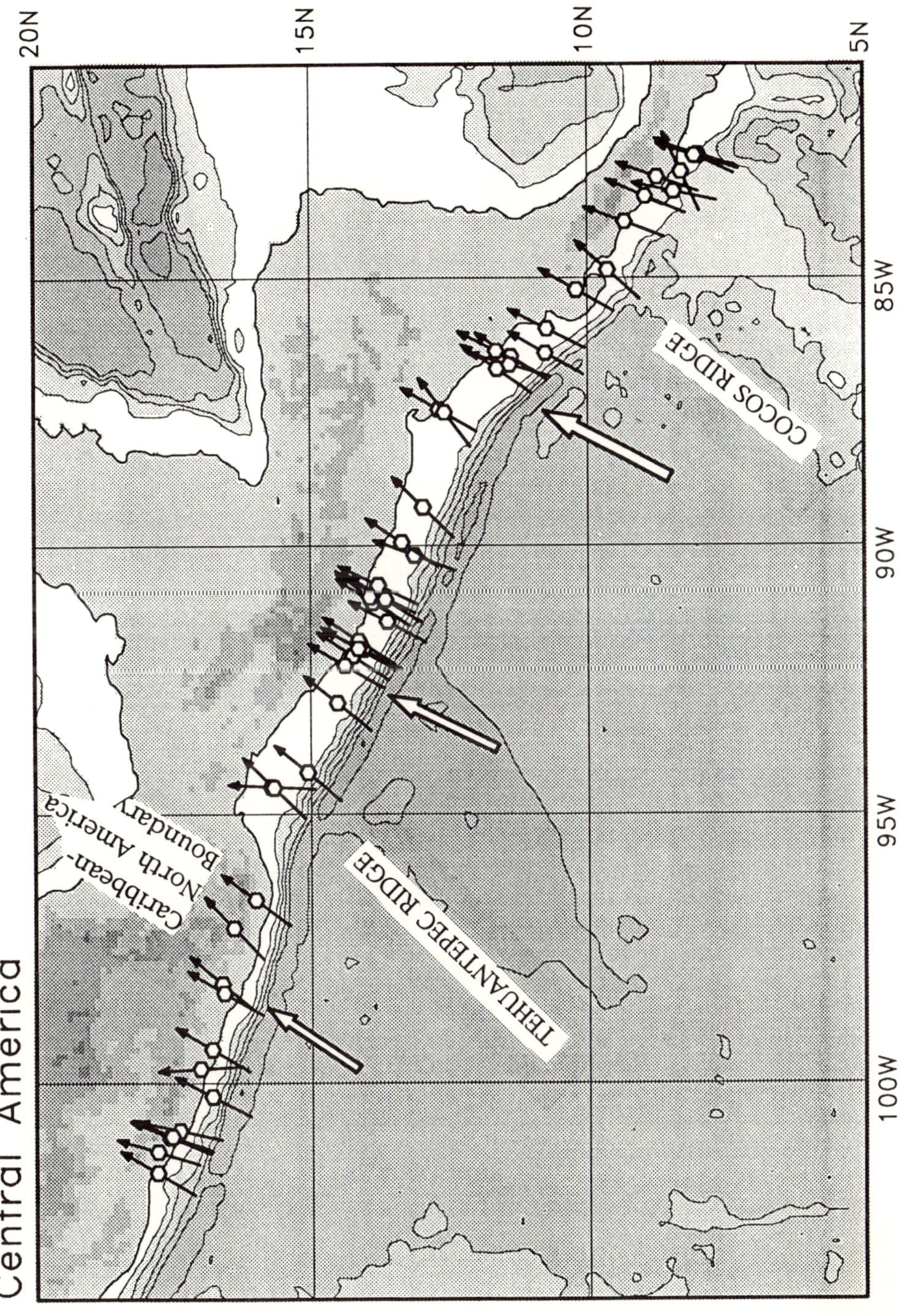

Figure 5(a)

Central America

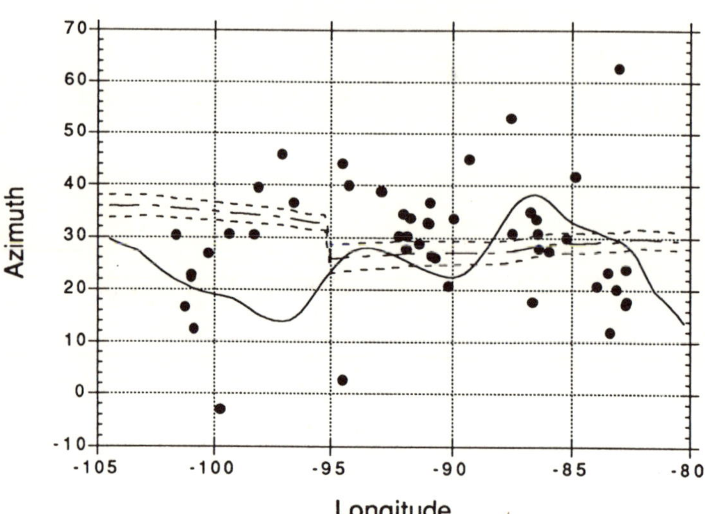

Figure 5(b)

Central America

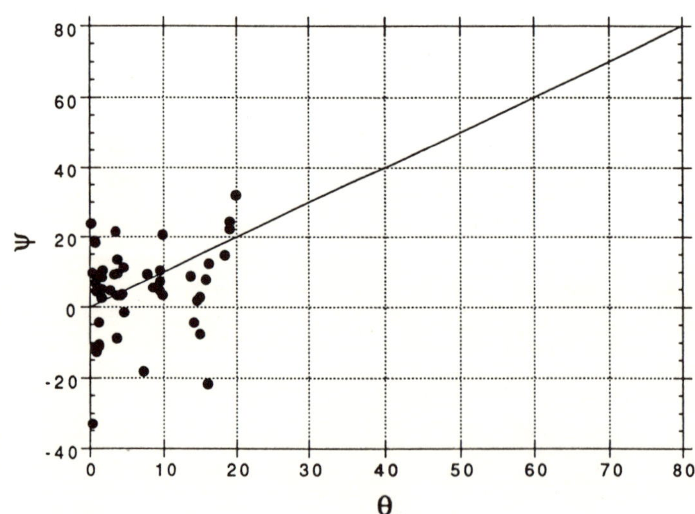

Figure 5(c)

Figure 5
Observations for Central America. See Figure 3 caption for explanation.

The Central American (Figure 5) and Japan and Kurile (Figure 6) subduction zones show a scatter in observed slip azimuths ψ which is on the same order as the observed range of obliquity θ ($<30°$). Hence, the observations are of limited utility in attempting to infer any relationship of ψ to θ.

There are relatively few slip azimuths available in the Harvard Catalog along Sumatra (Figure 7) and there also exist large uncertainties in the relative plate motion. The NUVEL-1 model predicts that the azimuth of convergence is about 25° along the Sumatran arc and reflects motion between the Australian and Eurasian plates. However, the NUVEL-1 model does not account for internal

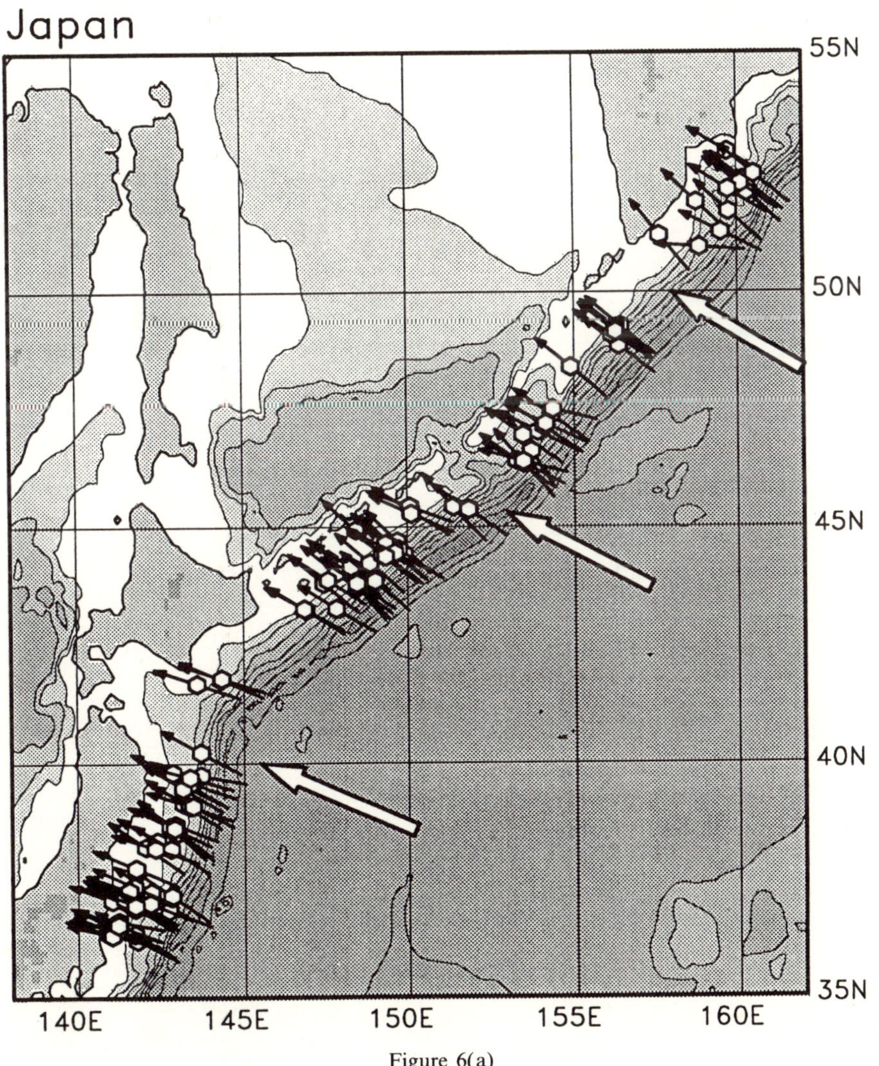

Figure 6(a)

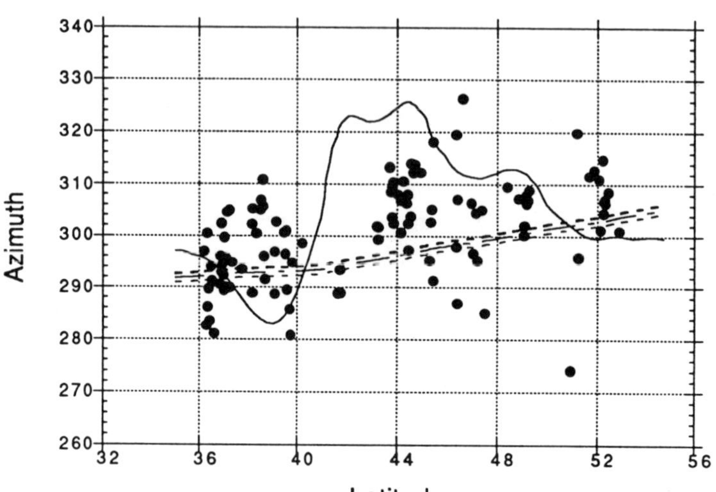

Figure 6(b)

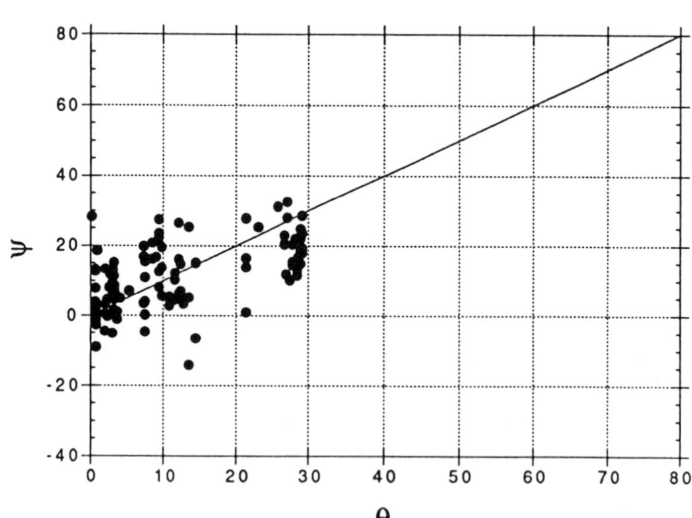

Figure 6(c)

Figure 6

Observations for Kurile and Japan. See Figure 3 caption for explanation.

deformation within Southeast Asia which is now well recognized (McCAFFREY, 1991); hence, the predicted motions likely do not reflect the actual relative motion across the Sumatran arc. McCAFFREY (1992) argued that such internal deformation is absent to the east of Sumatra and, on that basis, asserted that focal mechanisms east of Sumatra were representative of the plate motion direction of N3°E across the arc. However, because earthquake slip vectors are generally unreliable indicators of relative plate motion (e.g., DeMETS *et al.*, 1990), and plate motion directions inferred from earthquake slip directions do not provide an independent measure of plate motion on which to base determinations of the degree

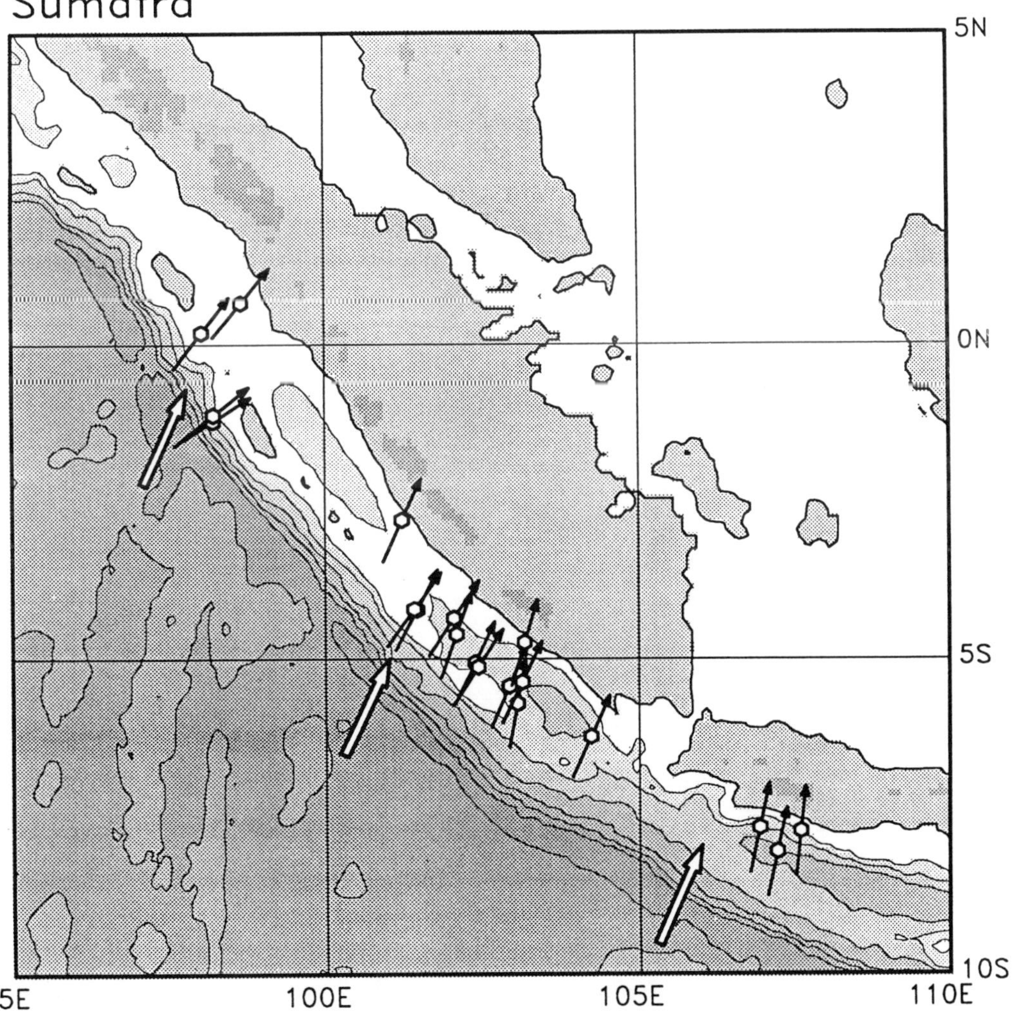

Figure 7(a)

Sumatra

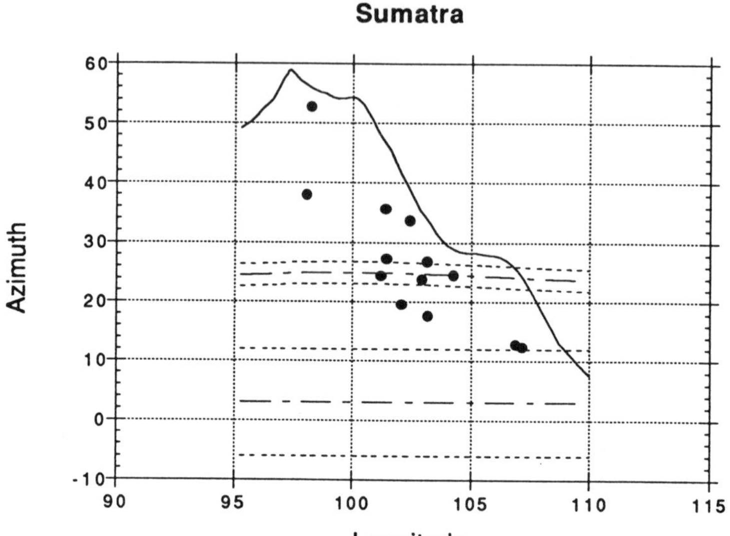

Figure 7(b)

Sumatra

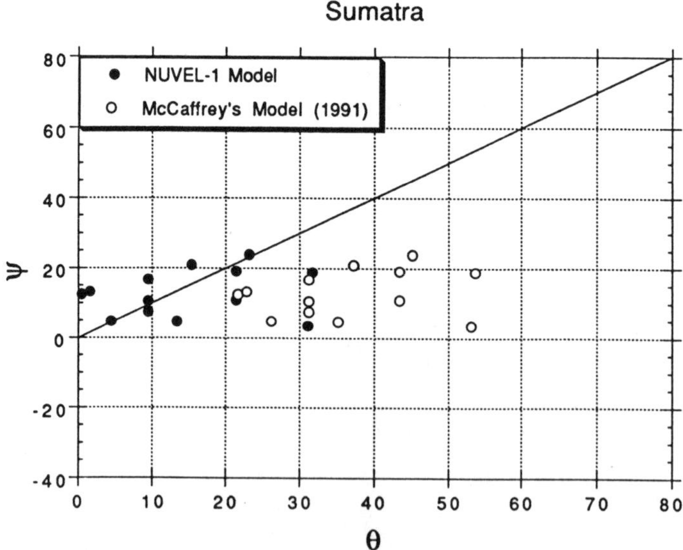

Figure 7(c)

Figure 7

Observations for Sumatra. See Figure 3 caption for explanation. The additional three lines on the bottom of figure represents McCaffrey's prediction of relative plate motion and the uncertainty for Sumatra (MCCAFFREY, 1991). Solid circles and open circles in Figure 3b correspond to slip partitioning angles obtained by using NUVEL-1 plate motion model and McCaffrey's plate motion prediction, respectively.

of partitioning, any estimate of the degree of partitioning along the Sumatran arc must be viewed with uncertainty and caution. For reference, the predicted motions of NUVEL-1 and MCCAFFREY (1991) are shown in Figures 7b and 7c. Given that the plate motion direction is nearly due north as asserted by MCCAFFREY (1991), values of ψ fall well below the line defined by $\psi = \theta$ and do not appear to increase as a function of θ as observed for the Aleutians and possibly South America.

Arcs with Back-arc Spreading

Each of the Tonga and Kermadec (Figure 8), the Mariana and Izu-Bonin (Figure 9), and New Hebrides Island arcs (Figure 10) are characterized by varying

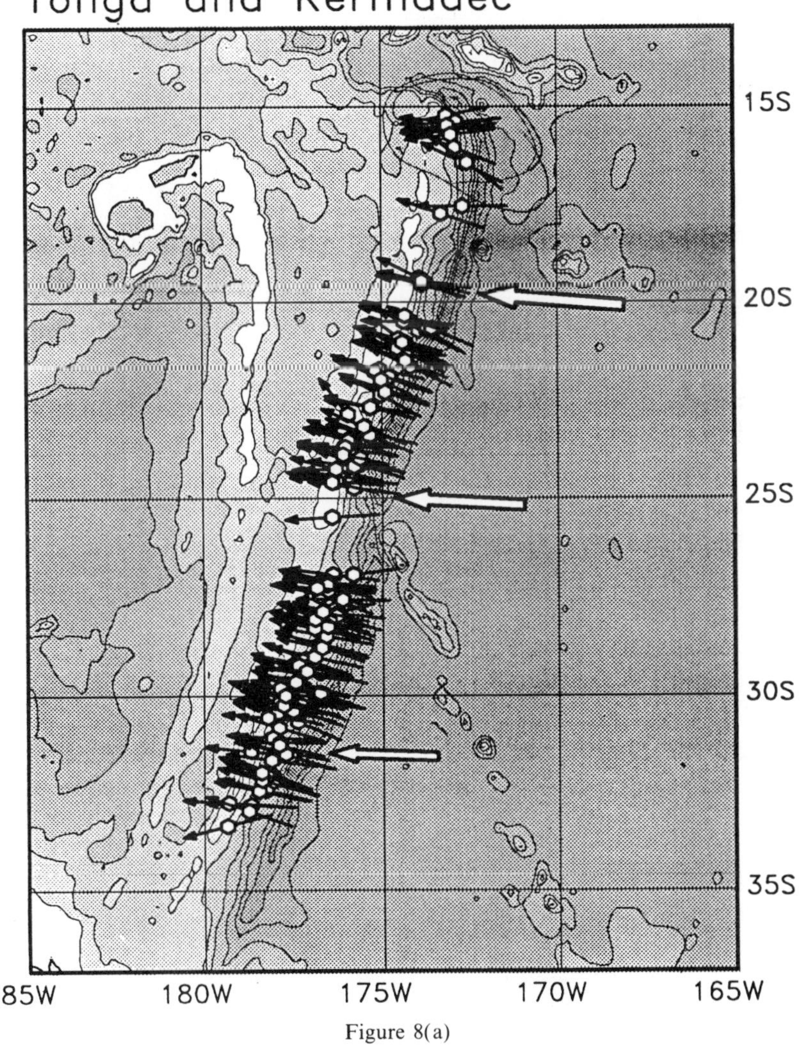

Figure 8(a)

Tonga and Kermadec

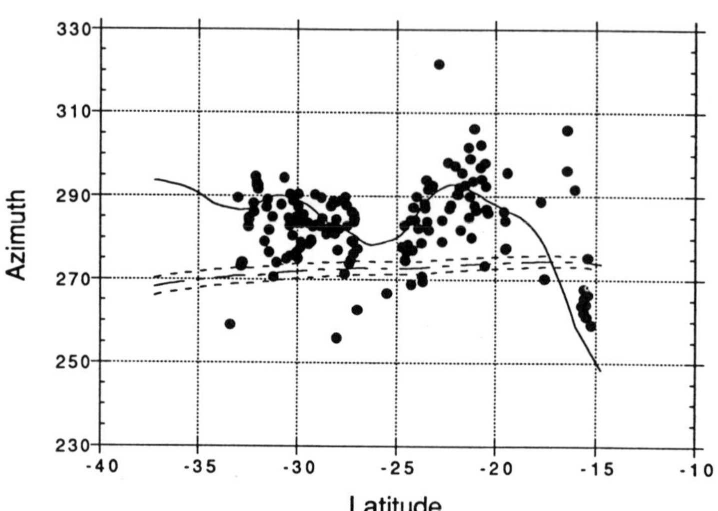

Figure 8(b)

Tonga and Kermadec

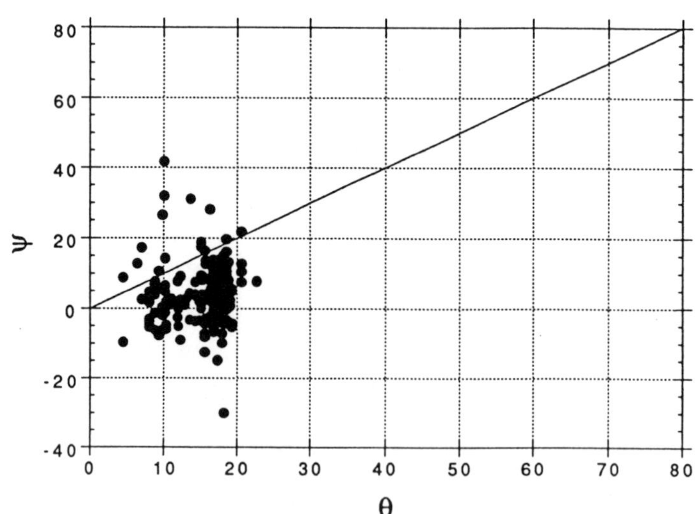

Figure 8(c)

Figure 8
Observations for Tonga and Kermadec. See Figure 3 caption for explanation.

amounts of back-arc spreading (JARRARD, 1986). Along the Tonga and Kermadec arcs, little insight is to be gained regarding any relation between ψ and θ because of the large scatter in observed slip azimuths ψ and the small range in obliquity θ observed along the arcs (Figure 8c).

The range in obliquity values along the Marianas and Izu-Bonin arcs is much larger, ranging between about 15° and 70° (Figure 9); and the few slip azimuths of earthquakes clearly rotate to the trench normal, indicating a large degree of partitioning. Similar to Sumatra, earthquake slip vectors along the Izu-Bonin were used by SENO (1987) to determine the relative motion of the Philippine Sea plate and, hence, determination of the degree of partitioning might be questioned along

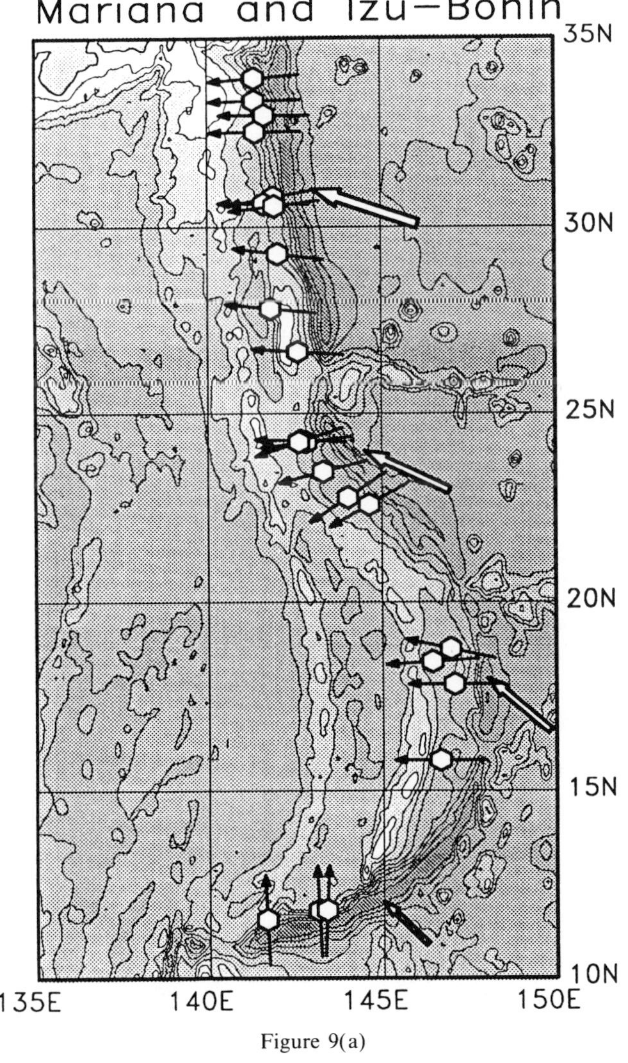

Figure 9(a)

G. Yu *et al.*

Mariana and Izu-Bonin

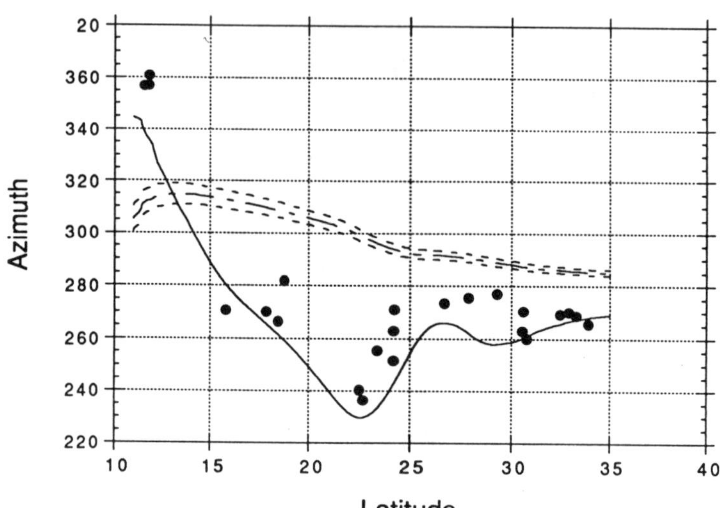

Figure 9(b)

Mariana and Izu-Bonin

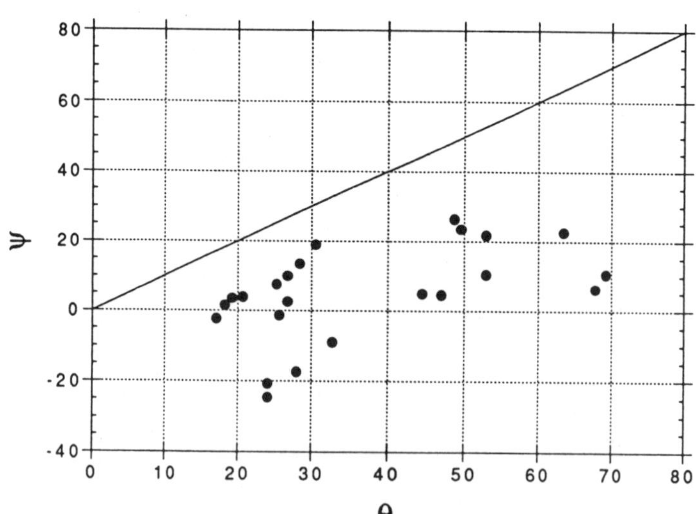

Figure 9(c)

Figure 9
Observations for Mariana and Izu-Bonin arcs. See Figure 3 caption for explanation. Plate motion vectors from SENO *et al.* (1987).

the Izu-Bonin and Marianas arcs. However, because earthquake slip vectors are rotated virtually to the trench normal vector along the Marianas arc, which is characterized by a 50° range of obliquity values (Figure 9), the result that slip partitioning is large along the Marianas should not be affected by uncertainties in the plate motion vector used in the analysis.

The slip azimuths of earthquakes along the New Hebrides also clearly trend parallel to the trench normal over a large range of obliquity values (Figure 10a), indicating that partitioning along the New Hebrides is virtually complete, although the apparent systematic rotation of slip azimuths past the trench normal is

New Hebrides

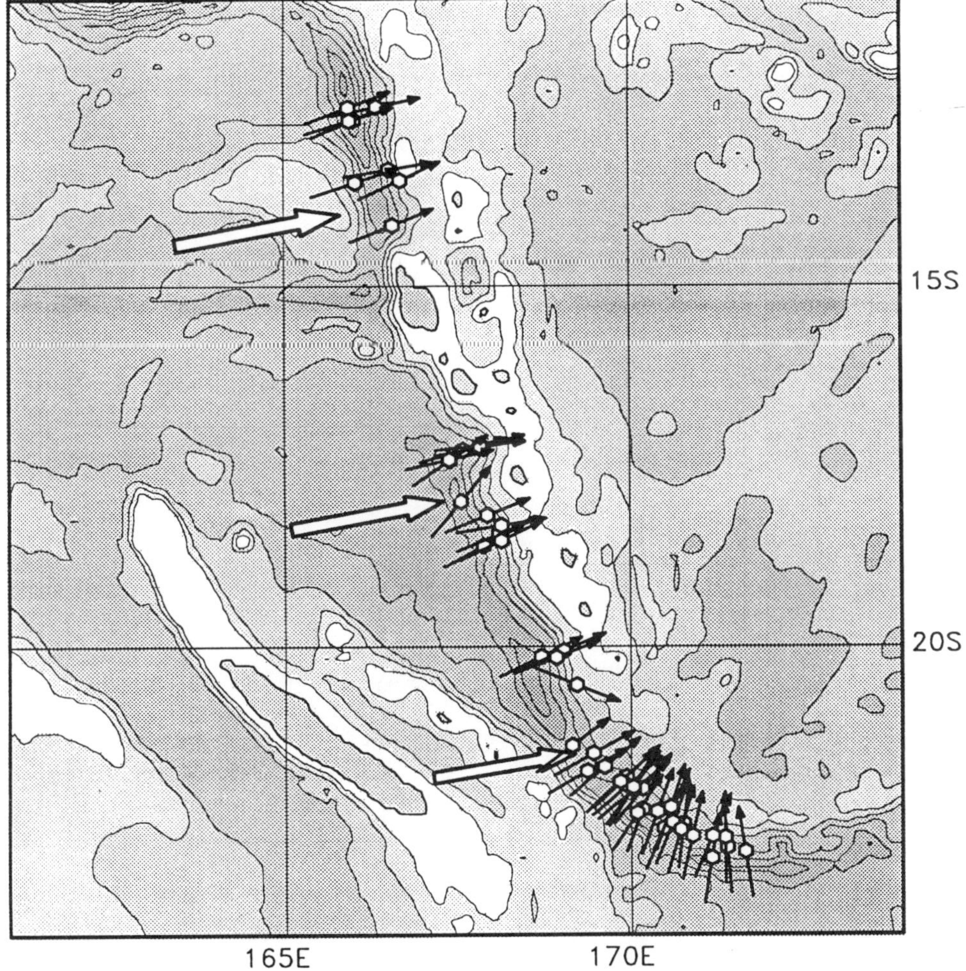

Figure 10(a)

New Hebrides

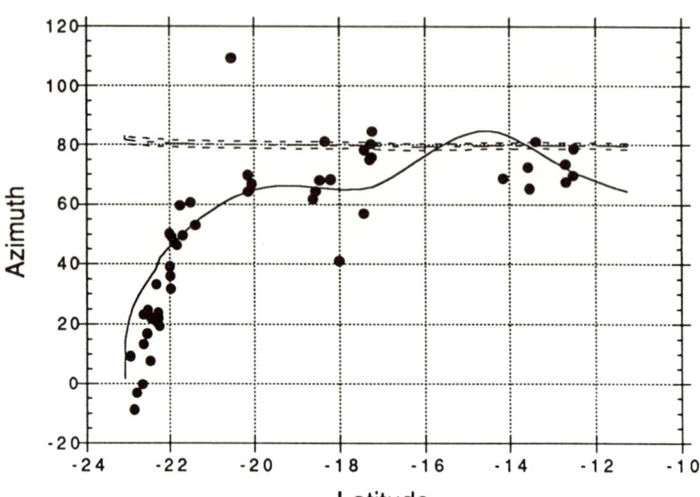

Figure 10(b)

New Hebrides

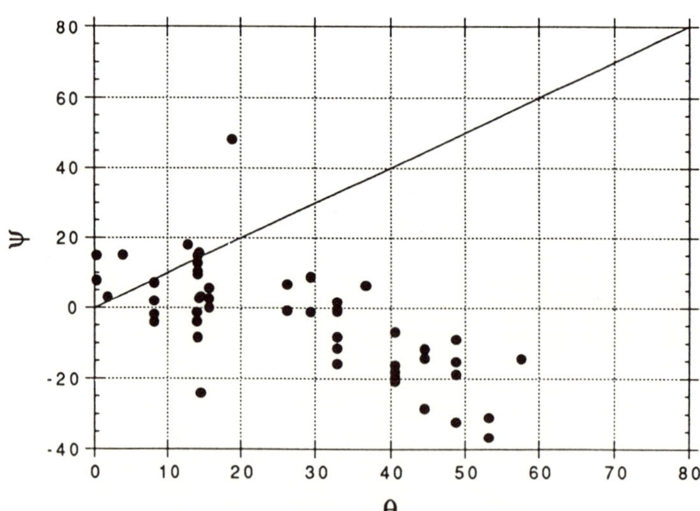

Figure 10(c)

Figure 10

Observations for New Hebrides. See Figure 3 caption for explanation.

enigmatic (Figures 10a and 10b). Portions of the New Hebrides characterized by values of obliquity θ greater than 30° are limited to the portion of arc south of latitude 22°S where there exists extreme curvature of the arc (Figures 10a and 10b). Perhaps the apparent systematic rotation of earthquake slip vectors past the trench normal directions is a result of the extreme contortion of the subducting plate which must accompany the severe bend in the trench. Nonetheless, it is clear that partitioning tends toward completeness along the New Hebrides.

Discussion

Synopsis

A synopsis of the relationship of ψ to θ between the various arcs is provided in Figure 11, a plot of ψ versus θ for all earthquakes along each arc. Open and solid

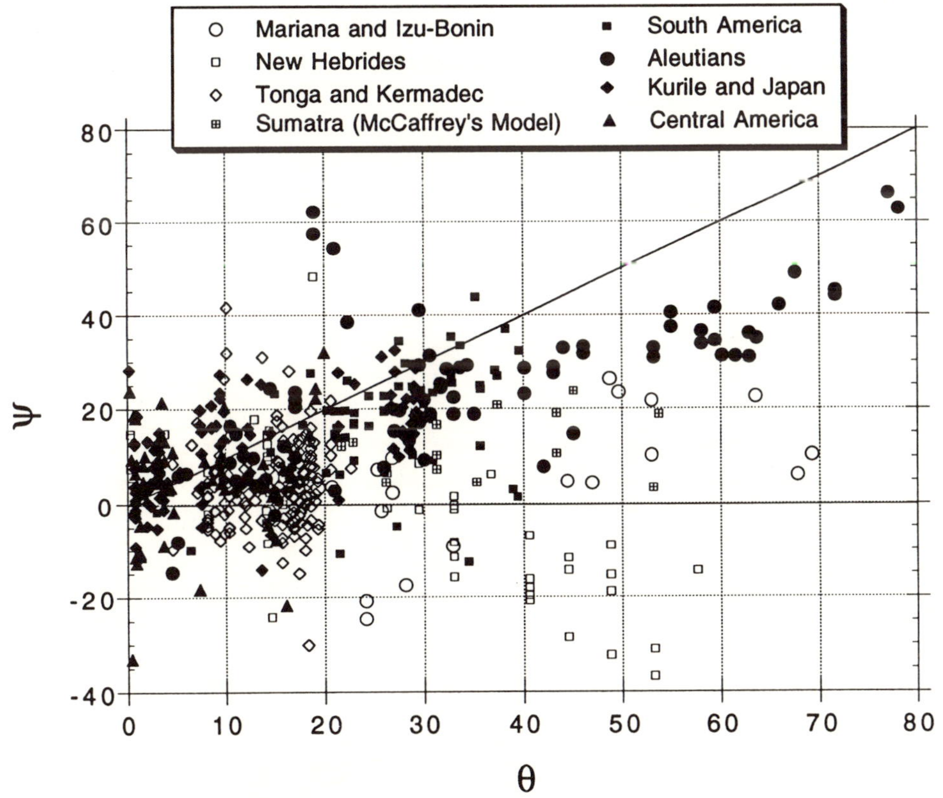

Figure 11

A plot of slip partitioning angle ψ versus θ for all earthquakes in this study, delineated according to the subduction zone along which they occur. Open and closed symbols represent boundaries with and without back-arc spreading, respectively. Data points for Sumatra calculated using plate motion model of MCCAFFREY (1992).

symbols in the figure correspond to earthquakes in those arcs which do and do not show evidence of back-arc spreading, repectively. Sumatra is absent of back-arc spreading but, because of the uncertainty attendant to determining the plate motion direction for the arc, the data points are shaded. The line $\psi = 0$ corresponds to total partitioning (slip vector perpendicular to the trench), and the line $\psi = \theta$ indicates no partitioning (slip vector parallel to the plate motion direction). Although there exists considerable scatter in the observed azimuths of earthquake slip vectors along each arc examined, the data presented in Figure 11 provide a basis to suggest that first-order differences in the degree of partitioning do exist between various arcs. Slip partitioning appears to be consistently large along the Marianas and New Hebrides arcs, each of which show back-arc extension. A similar degree of partitioning may characterize Sumatra where strike-slip faulting occurs in the overriding plate, but again the result should be viewed with some caution because earthquake slip vectors along the arc have also been the basis to determine the relative plate motion vector used in defining the degree of partitioning. The least amount of partitioning observed is for the Alaskan-Aleutian arc, where it is additionally seen that the slip partitioning angle ψ along the Alaskan-Aleutian arc appears to systematically increase as a function of the obliquity of convergence. Within the scatter of the data, a similar increase in ψ as a function of θ is not indicated for the New Hebrides, Marianas or Sumatran arcs. The scatter in slip vectors, coupled with the small range of obliquity values, generally precludes the ability to make comparisons to the other arcs.

Subduction Zones with Back-arc Spreading

In the case of those subduction zones characterized by back-arc spreading, the question arises whether or not the amount and direction of back-arc spreading are sufficient to entirely account for the rotation of earthquake slip vectors toward the trench normal. Figure 1a represents the case where back-arc spreading is oriented perpendicular to the trench. The relative plate motion vector V_{PMV} is accounted for by the sum of the back-arc spreading vector V_{NSV} and the earthquake slip vector V_{ESV}, which requires the earthquake slip vector to rotate away from the plate motion vector V_{PMV} and toward the trench normal vector V_{TNV}. The amount of back-arc spreading needed to provide for slip partitioning is then written as

$$V_{NSV} = \frac{V_{PMV}\sin(\theta - \psi)}{\sin\psi}, \quad \psi \neq 0 \tag{1}$$

or, when back-arc spreading is oriented at an oblique angle α with respect to the trench normal vector (Figure 1b), the amount of back-arc spreading needed to allow for slip partitioning is

$$V_{OSV} = \frac{V_{PMV}\sin(\theta - \psi)}{\sin(\psi + \alpha)}, \quad \psi \neq -\alpha. \tag{2}$$

For the oblique spreading case, and within the range of θ and ψ values observed, a relatively smaller value of back-arc spreading is needed to produce a given rotation of the earthquake slip vectors, as compared to spreading parallel to the trench normal vector (Figure 1a). The two models in Figures 1a and 1b, although certainly not representing the true complexity of the partitioning process, provide a first-order framework to examine whether or not observed back-arc spreading rates are sufficient to account for observed values of partitioning.

For example, HUSSONG and UYEDA (1982) place the rate and direction of spreading at 4.3 cm/yr and 60°, respectively, at latitude 18°N along the Marianas arc. SENO et al. (1987) places the rate and direction of convergence across the subduction zone at 4.5 cm/yr and 310°, respectively. Our best estimate of the azimuths of the trench normal and earthquake slip vector at the same latitude are 266° and 271°, respectively. Within the framework of Figure 1b, we can then define ψ, θ, and α to equal 5°, 44°, and 26°, respectively. In turn, we can use equation (2) to determine the amount of oblique spreading needed to account for the observed rotation of earthquake slip vectors to the trench normal to equal 5.5 cm/yr. The 5.5 cm/yr rate is greater than the 4.3 cm/yr rate observed by HUSSONG and UYEDA (1982), implying that observed rates of spreading are not sufficient to fully provide for the degree of slip partitioning implied by the rotation of earthquake slip vectors along the arc.

Behind the New Hebrides, seafloor magnetic lineations mark a well-defined spreading center which strikes north between 18°S and 21°S latitude and is interpreted to have a spreading rate between 6.8 cm/yr and 8.2 cm/yr (AUZENDE et al., 1988, CHASE, 1971; MAILLET et al., 1989; MALAHOFF et al., 1982). However, the largest values of slip partitioning are observed to the south of 21°S where back-arc spreading is not recognized immediately behind the arc. Although the New Hebrides arc is characterized by back-arc spreading, the extreme rotation of earthquake slip vectors toward the trench normal along the southern end of the arc cannot directly be attributed to the occurrence of back-arc spreading. Indeed, it has previously been pointed out, on the basis of distributed seismicity and focal mechanisms, that the region behind the southern New Hebrides arc is characterized by a diffuse zone of left-lateral shear on east to northeast striking planes (HAMBURGER et al., 1988; LOUAT and PELLETIER, 1989), which is in the correct sense to accommodate the oblique component of convergence across this section of the subduction zone. It appears that it is shearing behind the trench and not back-arc spreading which is accommodating the large degree of partitioning observed along the southern portion of the New Hebrides (LOUAT and PELLETIER, 1989).

Subduction Zones without Back-arc Spreading

For the case where the oblique component of plate motion is not accommodated by some form of extension behind the trench, the oblique component of motion not accommodated by slip along the thrust interface must be accommodated by strike-slip motion behind the trench (Figure 1c). The amount of the strike-slip motion needed to accommodate the slip partitioning can be expressed (JONES and WESNOUSKY, 1992) as

$$V_{SSV} = \frac{V_{PMV} \sin(\theta - \psi)}{\cos \psi}. \tag{3}$$

Implicit to equation (1), as well as equations (2) and (3), is that the forearc behaves as a rigid sliver. The actual mechanism for partitioning is likely more complicated than these idealized models of Figure 1. For example, GEIST and SCHOLL (1992) have recently modeled the deformation of the Aleutian arc as occurring diffusely across the arc using a thin viscous sheet model. The lack of clearly identified major strike-slip faults behind several subduction zones which exhibit slip partitioning suggests that this may be an appropriate description of the deformation. In these cases we interpret the inferred strike-slip motions as representing the along-arc shear strain integrated across the arc and back-arc. Additionally, it is clear for the case of strike-slip faulting behind the trench that changes in either θ or ψ (equation (3)) along strike of a trench will result in a stretching or contraction of the forearc (EKSTRÖM and ENGDAHL, 1989; McCAFFREY, 1991, 1992). In the case of the southern New Hebrides, relative plate motion is about 8 cm/yr, obliquity θ of convergence ranges between about 35° to 60°, and slip azimuths ψ are about 0° between latitude 22°S and 23°S. Input of these values into equation (3) then implies that about 5 to 7 cm/yr of strike-slip motion behind the trench is needed to accommodate the oblique component of plate motion.

The mechanism of partitioning of oblique convergence between strike-slip faulting behind the trench and thrusting along the plate interface was initially recognized and explored by FITCH (1972). More recently, BECK (1989, 1991) and JONES and WESNOUSKY (1992) assumed that the relative amount and direction of slip on a paired oblique thrust and strike-slip system (Figure 1c) would be that which yields the minimum work per unit of convergence. The result of these efforts was to define a relationship between slip partitioning and the convergence parameters of obliquity, the dip angle of subduction, and the ratio of resistance to slip on the strike-slip fault to resistance of slip on the subduction surface. McCAFFREY (1992) has argued that paired systems of strike-slip and thrust faults along subduction zones do not extend to the same depth and used force-balance constraints to modify the approach of BECK (1989, 1991) and JONES and WESNOUSKY (1992).

McCAFFREY (1992) inferred that the shear force that drives the slip on the strike-slip fault F_s is derived from the horizontal component of shear force on the thrust fault F_x. Thus, for small values of obliquity, the shear force on the strike-slip fault F_s is not large enough to cause it to slip, and slip will occur on the thrust fault parallel to the plate convergence vectors so that the slip partitioning angle equals obliquity angle ($\psi = \theta$, no partitioning). Larger values of obliquity θ will result in a larger horizontal component of shear force F_x, so shear force F_s on the strike-slip fault also increases. At a critical obliquity angle, F_x equals F_s and the strike-slip fault becomes active. MCCAFFREY defines this critical value of the obliquity angle as $\psi_{\max}$, and obtained the following relationship

$$R_f = \sin \psi_{\max} = \frac{Z_s \tau_s \sin \Delta}{Z_t \tau_t} \tag{4}$$

when assuming that the shear forces acting on the coupled fault system are in equilibrium, where R_f is the ratio of the shear forces resisting motion on the two faults, Z_s and Z_t are the depths to the base of the strike-slip fault and the thrust fault respectively, τ_s and τ_t are average shear stresses on the strike-slip fault and the thrust fault respectively, and Δ is the dip angle of the thrust fault. MCCAFFREY argues that ψ cannot increase further than $\psi_{\max}$ even as the obliquity θ continues to increase. Thus, the slip partitioning angle ψ will equal obliquity angle θ (slope of 1) until reaching a critical value of $\theta = \psi_{\max}$, and will then remain at angle $\psi_{\max}$ for values of obliquity $\theta > \psi_{\max}$ (Figure 12). For the purpose of the following discussion, we will limit attention to the formulation of MCCAFFREY (1992) and simply note the same inferences are yielded by the formulations of BECK (1989, 1991) and JONES and WESNOUSKY (1992).

For those boundaries lacking back-arc spreading, it is only along the Aleutians that there exists a wide variation in the value of obliquity θ along strike. MCCAFFREY (1992) earlier used the Aleutian data set (Figure 3c) to support the behavior implied by equation (4) and illustrated in Figure 12. Indeed, when including all the events in Figure 3c (both solid and open symbols), it is permissible to describe the relationship between ψ and θ in the form of Figure 12. However, when limiting

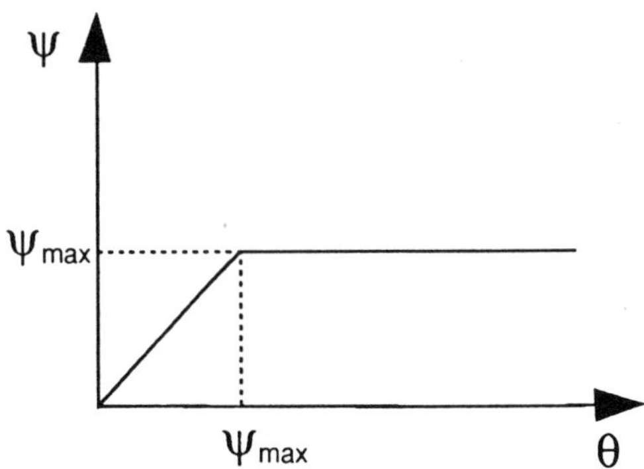

Figure 12

A schematic plot of slip partitioning angle ψ versus obliquity of convergence θ which is expected for simple model of slip partitioning described by equation (4): $\psi = \theta$ at small obliquity and ψ remains constant when obliquity θ reaches a critical angle $\psi_{\max}$.

attention to the events (solid symbols in Figure 3c) with strike parallel ($\pm 20°$) to the trench, dips $<40°$, and larger seismic moments ($>1.5 \times 10^{24}$ dyne-cm), it appears more convincing that the relationship between ψ and θ is linear and of slope less than 1. The observations do not require that the model predictions (equation (4) and Figure 12) are incorrect but, rather, may imply that either (1) the ratio of shear strengths τ_s/τ_t, (2) the ratio of the depth of the strike-slip to thrust fault Z_s/Z_t, (3) the dip of the thrust fault, or (4) material properties vary systematically in a manner to increase the value of $\psi_{\max}$ along strike, of which each idea is briefly considered below.

In a separate study of slip partitioning using well-defined focal mechanisms of historical earthquakes along the San Andreas system, JONES and WESNOUSKY (1992) showed that the ratio of τ_s over τ_t is a decreasing function of the obliquity θ of convergence, which is exactly the opposite in sense required to explain the linear increase in ψ observed for the Alaskan-Aleutian arc (Figure 3c). Hence, within the context of equation (4), it seems unreasonable to attribute the observed increase in ψ to be the direct result of increasing obliquity θ as one progresses from east to west along the Alaskan-Aleutian arc. In contrast, cross sections and estimates of the dip of the Benioff zone (measured in the depth range between 0 and 60 km depth) as reported by JARRARD (1986) suggest an increase in dip from about 7 to 25 degrees from Alaska to the Central Aleutians, which is in the correct sense to explain the increase in ψ from east to west along the arc. Similarly, one might expect that the down-dip width of the thrust interface subject to brittle behavior decreases as obliquity of convergence increases because the down-dip component of convergence and, hence, the downward deflection of geotherms would correspondingly decrease (e.g., MCKENZIE, 1970). Such an effect might result in increasing the ratio of Z_s/Z_t and, hence, ψ as obliquity increases westward along the Aleutians, but we are not aware of any firm data verifying such an idea in the Aleutians.

It has also been suggested that seismic coupling along the Alaskan-Aleutian thrust interface decreases westward along the arc as a function of decreasing sediment supply to the trench westward from the Gulf of Alaska (RUFF and KANAMORI, 1983). The speculative suggestion may also work toward explaining the observed increase in ψ westward along the Alaskan-Aleutian arc as resulting from a systematic variation in material properties, given that seismic coupling is directly proportional to the shear strength τ_t of the thrust as described by equation (4).

Conclusion

The crux of our observations is summarized in Figure 11. To first order, we observe that the Marianas and New Hebrides arcs, each of which are characterized by back-arc spreading, tend to show the greatest degree of partitioning. However, the rates, directions and location of back-arc spreading are not sufficient to account for the observed rotation of earthquake slip vectors along the arcs, which implies that some portion of the oblique component of subduction is also accommodated by shearing within the overriding plate. Among those arcs lacking back-arc spreading, it is only along the Alaskan-Aleutian and Sumatran arcs where there exists a large variation along strike in the obliquity of predicted plate motion. The degree of partitioning along Sumatra approaches that observed along the Marianas, but is viewed with caution because the result is based on a relative plate motion vector also determined from earthquake slip vectors along the arc. In the case of the Alaskan-Aleutian arc, the degree of partitioning is relatively less and rotation of slip vectors toward the trench normal appears to increase as a function of the obliquity of convergence. Assuming that partitioning in the Alaskan-Aleutian arc is accommodated by strike-slip faulting within the upper plate, it is suggested that the positive relationship between obliquity of convergence and the rotation of earthquake slip vectors to the trench normal may reflect either (1) an increase in the ratio Z_s/Z_t of the depth of the strike-slip to thrust fault westward along the arc, (2) an increase in the dip of the subduction thrust westward along the arc, or (3) a decrease in the strength of the subduction thrust westward along the arc.

Acknowledgements

The authors thank Craig H. Jones for his time and thought during the course of this research, and the extremely critical and constructive review comments of an earlier version of the manuscript by Robert McCaffrey. Useful comments were also received from Eric Geist and M. Beck. The research was supported by National Science Foundation Grant EAR–8915833. Center for Neotectonics Contribution No. 2.

REFERENCES

AUZENDE, J. M., LAFOY, Y., and MARSSET, B. (1988), *Recent Geodynamic Evolution of the North Fiji Basin (Southwest Pacific)*, Geology *16*, 925–929.

BECK, M. (1983), *On the Mechanism of Tectonic Transport in Zones of Oblique Subduction*, Tectonophysics *93*, 1–11.

BECK, M., *Block rotations in continental crust: Examples from western North America, rotations and continental deformations* (eds. C. Kissel and C. Laj) (Kluwer Academic Publishers, 1989) pp. 1–16.

BECK, M. (1991), *Coastwise Transport Reconsidered: Lateral Displacements in Oblique Subduction Zones, and Tectonic Consequences*, Phys. Earth Planet. Int. *68*, 1–8.

CHASE, C. G. (1971), *Tectonic History of the Fiji Plateau*, Geol. Soc. Am. Bull. *82*, 3087–3110.

DEMETS, C., GORDON, R. G., ARGUS, D. F., and STEIN, S. (1990), *Current Plate Motions*, Geophys. J. Int. *101*, 425–478.

DZIEWONSKI, A. M., EKSTRÖM, G., WOODHOUSE, J. H., and ZWART, G. (1991), *Centroid-moment Tensor Solutions for October–December 1990*, Phys. Earth Planet. Int. *68*, 201–214.

ESTRÖM, G., and ENGDAHL, E. R. (1989), *Earthquake Source Parameters and Stress Distribution in the Adak Island Region of the Central Aleutian Islands, Alaska*, J. Geophys. Res. *94*, 15,499–15,519.

FITCH, T. J. (1972), *Plate Convergence, Transcurrent Faults and Internal Deformation Adjacent to Southeast Asia and the Western Pacific*, J. Geophys. Res. *77*, 4432–4460.

GEIST, E. L., and SCHOLL, D. W. (1992), *Application of Continuum Models to Deformation of the Aleutian Island Arc*, J. Geophys. Res. *97*, 4953–4967.

HAMBURGER, M. W., EVERINGHAM, I. B., and ISACKS, B. L. (1988), *Active Tectonism within the Fiji Platform, Southwest Pacific*, Geology *16*, 237–241.

HUSSONG, D. M., and UYEDA, S. (1982), *Tectonic Processes and the History of the Mariana Arc: A Synthesis of the Results of Deep Sea Drilling Leg 60*, Initial Rep. Deep Sea Drill. Proj. *60*, 909–929.

JARRARD, R. D. (1986), *Relations among Subduction Parameters*, Reviews of Geophysics *24* (2), 217–284.

JONES, C. H., and WESNOUSKY, S. G. (1992), *Variations in Strength and Slip Rate along the San Andreas Fault System*, Science *256*, 83–86.

LOUAT, R., and PELLETIER, B. (1989), *Seismotectonics and Present-day Relative Plate Motions in the New Hebrides–North Fiji Basin Region*, Tectonophysics *167*, 41–55.

MAILLET, P., MONZIER, M., EISSEN, J. P., and LOUAT, R. (1989), *Geodynamics of an Arc-ridge Junction: The New Hebrides Arc/North Fiji Basin Case*, Tectonophysics *165*, 251–268.

MALAHOFF, A., FEDEN, R. H., and FLEMING, H. F. (1982), *Magnetic Anomalies and Tectonic Fabric of Marginal Basins North of New Zealand*, J. Geophys. Res. *87*, 4109–4125.

MCCAFFREY, R. (1990), *Oblique Plate Convergence, Slip Vectors, and Forearc Deformation*, EOS Trans., AGU *71*, 1590.

MCCAFFREY, R. (1991), *Slip Vectors and Stretching of the Sumatran Forearc*, Geology *19*, 881–884.

MCCAFFREY, R. (1992), *Oblique Plate Convergence, Slip Vectors, and Forearc Deformation*, J. Geophys. Res. *97*, 8905–8915.

MCKENZIE, D. P. (1970), *Temperature and Potential Temperature beneath Island Arcs*, Tectonophysics *10*, 357–366.

RUFF, L., and KANAMORI, H. (1983), *Seismic Coupling and Uncoupling at Subduction Zones*, Tectonophysics *99*, 99–117.

SENO, T., MORIYAMAM, T., STEIN, S., WOODS, D. F., DeMETS, C., ARGUS, D., and GORDON, R. (1987), *Redetermination of the Philippine Sea Plate Motion*, EOS Trans., AGU *68*, 1474.

(Received January 15, 1992, accepted January 18, 1993)

PAGEOPH, Vol. 140, No. 2 (1993)

0033–4553/93/020211–45$1.50 + 0.20/0

Quantitative Estimates of Interplate Coupling Inferred from Outer Rise Earthquakes

XINPING LIU[1] and KAREN C. MCNALLY[1]

Abstract—Interplate coupling plays an important role in the seismogenesis of great interplate earthquakes at subduction zones. The spatial and temporal variations of such coupling control the patterns of subduction zone seismicity. We calculate stresses in the outer rise based on a model of oceanic plate bending and coupling at the interplate contact, to quantitatively estimate the degree of interplate coupling for the Tonga, New Hebrides, Kurile, Kamchatka, and Marianas subduction zones. Depths and focal mechanisms of outer rise earthquakes are used to constrain the stress models. We perform waveform modeling of body waves from the GDSN network to obtain reliable focal depth estimates for 24 outer rise earthquakes. A propagator matrix technique is used to calculate outer rise stresses in a bending 2-D elastic plate floating on a weak mantle. The modeling of normal and tangential loads simulates the total vertical and shear forces acting on the subducting plate. We estimate the interplate coupling by searching for an optimal tangential load at the plate interface that causes the corresponding stress regime within the plate to best fit the earthquake mechanisms in depth and location.

We find the estimated mean tangential load $\bar{f}_x$ over 125–200 km width ranging between 166 and 671 bars for Tonga, the New Hebrides, the Kuriles, and Kamchatka. This magnitude of the coupling stress is generally compatible with the predicted shear stress at the plate contact from thermal-mechanical plate models by MOLNAR and ENGLAND (1990), and VAN DEN BUEKEL and WORTEL (1988). The estimated tectonic coupling, F_{tc}, is on the order of $10^{12}–10^{13}$ N/m for all the subduction zones. F_{tc} for Tonga and New Hebrides is about twice as high as in the Kurile and Kamchatka arcs. The corresponding earthquake coupling force F_{ec} appears to be 1–10% of the tectonic coupling from our estimates. There seems to be no definitive correlation of the degree of seismic coupling with the estimated tectonic coupling. We find that outer rise earthquakes in the Marianas can be modeled using zero tangential load.

Key words: Interplate coupling, outer rise earthquakes, stress modeling, subduction zones.

Introduction

Two approaches are often adopted for plate tectonic studies. One is to study earthquakes occurring at the plate boundaries, which helps us to identify the shape and extent of a plate, to detect the degree of plate interaction, and to acquire information about the state of stress at the plate boundaries. Another approach is to construct mechanical models for plate motion, stresses, and evolution. Surface

---

[1] C. F. Richter Seismological Laboratory and Institute of Tectonics, University of California, Santa Cruz, Santa Cruz, CA 95064, U.S.A.

plate motion data sometimes are used to constrain the models, along with gravity and bathymetric data, and, occasionally, some earthquake and heat flow data as well. Results from the former approach often give only qualitative assessments of the parameters controlling the subduction dynamics, whereas quantitative results from the latter approach generally suffer from large uncertainties due to the need for many assumptions. Here, we combine the two approaches, using the information of earthquake depths and mechanisms to constrain the stress distribution in the plate, and attempt to obtain quantitative estimates of the parameters controlling the plate dynamics. In particular, we are interested in outer rise earthquakes and their relationship to subduction dynamics.

An important aspect of subduction zone dynamics is the degree of interplate coupling, which is related to the shear stress and coupling width at the interplate contact between the downgoing and overriding plates. Estimation of the shear stresses at the interplate contact is difficult, but the stress drops due to great underthrusting earthquakes can be estimated from seismological studies. RUFF and KANAMORI (1980, 1983) used the maximum moment magnitude of the characteristic underthrusting earthquakes in different regions to classify seismic coupling at subduction zones. They concluded that younger plates with faster convergence rates tend to have stronger seismic coupling. The Southern Chile subduction zone represents the end-member with strong seismic coupling. The Marianas, an old plate with a slower convergence rate, is an example of the other extreme, a weakly coupled subduction zone. Such general characterizations of seismic coupling are useful, but do not provide quantitative information about the degree of interplate coupling.

Outer rise earthquakes are oceanic intraplate events, occurring within the bending plate before it subducts into the trench. These earthquakes are rare compared with subduction zone interplate and intermediate depth intraplate seismicity. Approximately 20 outer rise earthquakes with magnitudes from 5 to 7 occur worldwide each year. The depths and focal mechanisms of outer rise earthquakes contain valuable information about the stress distribution with depth in the oceanic lithosphere near the trench. This distribution is related to the mechanical coupling at the interplate contact.

In general, the observation that tensional outer rise events tend to be shallower, while compressional outer rise events tend to be deeper, supports the association of these earthquakes with bending of the subducting plate (STAUDER, 1968a,b, 1973). However, variations in the degree of interplate coupling may modulate the stress in the outer rise from one subduction zone to another. Furthermore, temporal variations in interplate coupling related to great underthrusting earthquakes may superimpose time-dependent changes in the outer rise seismicity (WARD, 1983, 1984; CHRISTENSEN and RUFF, 1983, 1988; DMOWSKA et al., 1988; LAY et al., 1989). Spatial and temporal changes in stress inferred from outer rise earthquakes may thus provide constraints on the interplate coupling at subduction zones. The stress in the outer rise has been simulated to understand the mechanical behavior of the bending plate (CALDWELL *et al.*, 1976; CHEN and FORSYTH, 1978; TURCOTTE and MCADOO, 1978; CHAPPLE and FORSYTH, 1979; FORSYTH, 1982; TURCOTTE and SCHUBERT, 1982; WARD, 1984). A variety of plate models and rheologies have been used to match the observed bathymetric and gravity profiles near the ocean trenches, and they yield quite different stresses in the plate. It is evident that the problem of estimating the stresses in the oceanic lithosphere is highly underdetermined. Studying outer rise earthquakes can possibly better constrain such plate models and the stresses in the plate.

In this paper, we calculate stresses in the outer rise using a 2-D elastic bending plate model for the Tonga, New Hebrides, Kurile, Kamchatka, and Marianas subduction zones. We apply a tangential load at the interplate contact to simulate interplate coupling at each subduction zone; this load is constrained by the depths and focal mechanisms of the outer rise earthquakes. Our model estimates the degree of interplate coupling, represented by the tectonic coupling force F_{tc}, for each of the selected subduction zones. These estimates of F_{tc} provide useful information for understanding the environment of the large underthrusting earthquakes.

Earthquake Data and Focal Depth Determination

We select 24 outer rise earthquakes (listed in Table 1a) in the Tonga, New Hebrides, Kurile, Kamchatka, and Marianas trenches shown in Figure 1. The earthquakes were recorded by the Global Digital Seismographic Network (GDSN) between January, 1980 and August, 1986, with $m_b \geq 5.0$ We perform the stress modeling for these trenches because relatively numerous outer rise earthquakes occurred in each region with both tensional and compressional faulting mechanisms (except for the Marianas). Only an upper limit of the estimated tectonic coupling force can be obtained, based on the tensional events in the Marianas. We use the Harvard Centroid Moment Tensor (HCMT) focal mechanisms (DZIEWONSKI *et al.*, and references therein, 1988) and PDE locations of earthquakes in the initial selection of outer rise events.

We apply a teleseismic body-wave modeling technique (WARD, 1983) and the HCMT focal mechanisms to determine the depths of the outer rise earthquakes. The usefulness of outer rise earthquakes for this study depends on their depth precision. Routinely determined catalog depths are not sufficiently reliable for the purpose of inferring stress regimes at depth in the lithosphere. Furthermore, existing data compilations (locations, depths and focal mechanisms of outer rise earthquakes) from various workers are too heterogeneous for our work because of differences in the methodologies and data used to determine earthquake depths, which leads to a large scatter in the reported depths. Our objective here is to upgrade existing data to a higher level of relative precision and

Table 1a

Selected outer rise earthquakes

Trench	Event Num.	Date	Time	Latitude	Longitude	Catalog Depth (km)	m_b	M_s	Redetermined Depth (km)*	Depth Uncertainty −dh	+dh	Focal Mech.
Tonga	1	02/03/80	11:58:40	−17.649	−171.184	33.0	6.2	6.4	25.0	−5.	5.	C
	2	12/15/80	08:12:45	−17.593	−172.300	33.0	6.1	6.3	20.0	−8.	12.	C
	3	11/25/81	19:01:48	−15.246	−173.308	33.0	5.6	5.8	30.0	−10.	10.	T
	4	02/28/82	17:00:24	−21.698	−173.538	37.0	5.6	5.6	30.0	−8.	10.	C
	5	06/02/82	12:37:35	−18.083	−172.492	33.0	6.4	6.4	5.0	−5.	5.	T
	6	05/11/83	21:48:15	−21.432	−173.453	33.0	5.7	5.3	35.0	−8.	4.	C
	7	08/30/83	08:50:17	−16.708	−172.082	38.7	6.0	5.7	35.0	−10.	8.	C
New Hebrides	1	02/17/81	15:18:34	−21.743	169.377	30.0	5.6	6.7	15.0	−5.	12.	C
	2	11/16/81	13:53:19	−22.141	169.503	33.0	5.7	6.2	15.0	−7.	5.	T
	3	05/20/82	21:29:15	−20.283	168.220	38.0	5.9	5.8	15.0	−6.	4.	T
	4	10/21/85	02:36:11	−13.555	166.020	46.3	5.5	5.0	10.0	−5.	8.	C
	5	11/28/85	02:25:42	−14.012	166.235	33.0	6.1	7.0	15.0	−15.	15.	T
	6	12/16/85	08:04:6	−14.153	166.356	33.0	6.0	6.7	15.0	−15.	10.	T
Marianas	1	09/04/82	07:56:6	15.559	147.581	46.0	5.5	5.2	15.0	−10.	13.	T
	2	03/17/83	02:53:11	12.298	143.974	52.0	5.6	0.0	15.0	−10.	5.	T
	3	11/22/83	02:07:13	12.082	144.361	43.3	5.6	5.3	10.0	−10.	5.	T
	4	09/20/84	19:19:23	16.786	147.217	48.2	5.7	5.5	20.0	−10.	6.	T
Kurile	1	04/30/81	14:41:41	43.233	150.222	49.0	6.1	6.1	15.0	−9.	15.	T
	2	06/30/82	01:57:34	44.679	151.143	33.0	6.6	6.9	30.0	−10.	10.	C
	3	08/02/83	06:08:7	45.139	153.485	56.0	5.5	5.0	30.0	−15.	5.	C
Kamchatka	1	02/01/81	22:43:28	53.030	162.406	33.0	5.9	5.4	15.0	−5.	10.	T
	2	08/23/81	12:00:27	48.718	157.390	40.0	6.0	5.8	15.0	−5.	5.	C
	3	10/01/81	17:04:45	50.733	160.429	33.0	5.9	5.7	25.0	−15.	8.	T
	4	09/26/82	01:09:29	50.053	158.798	44.0	5.5	4.5	15.0	−6.	8.	T

* Earthquake focal depth is measured from the top of oceanic plates.

Table 1b

Outer rise earthquakes from CR's data, used in the combined data set

Trench	Event Num.	Date	Time	ISC Location		m_b	M_s	Focal Mech.	Depth	Depth References
				Lat.	Lon.					
Tonga	T2	11/12/67	10:36	−17.190	−171.980	5.6	—	C	42.0	CF
	T5	08/07/72	09:24	−16.660	−172.010	5.8	6.0	C	45.0	F
	T6	09/27/72	09:01	−16.470	−172.170	5.8	6.0	T	6.0	CF
	T10	04/02/77	07:15	−16.790	−172.020	6.4	7.6	C	50.0	G
	T14	06/17/78	15:11	−17.060	−172.280	6.5	7.0	T	11.0	G
	T15	11/13/79	20:43	−23.610	−174.850	6.4	6.6	T	10.0	G
	T18	09/01/81	09:29	−15.080	−173.120	6.5	7.5	T	20.0	DW83
	T23	07/08/83	10:05	−21.500	−173.370	5.4	4.8	C	49.0	D84A
	T25	03/22/84	14:13	−15.230	−172.190	5.4	5.2	T	10.0	D84C
	T26	06/29/84	04:28	−20.930	−173.270	5.4	—	C	51.0	D85A
New Hebrides	N1	01/22/64	23:59	−13.640	165.960	6.3	—	T	21.0	CI
	N2	09/12/66	11:29	−23.000	170.600	5.9	—	T	18.0	CI
	N6	09/09/82	16:40	−22.050	169.380	5.5	5.0	T	10.0	D83A
Kurile	K4	04/05/65	13:52	44.510	150.900	5.6	—	T	25.0	F
	K7	09/09/71	23:01	44.340	150.850	6.0	5.9	T	15.0	F

References are CF, CHEN and FORSYTH (1978); CI, CHINN and ISACKS (1983); D83A, DZIEWONSKI *et al.* (1983a); D84A, DZIEWONSKI *et al.* (1984a); D84C, DZIEWONSKI *et al.* (1984c); D85A, DZIEWONSKI *et al.* (1985a); DW83, DZIEWONSKI and WOODHOUSE (1983); F, FORSYTH (1982); G, GIARDINI *et al.* (1985).

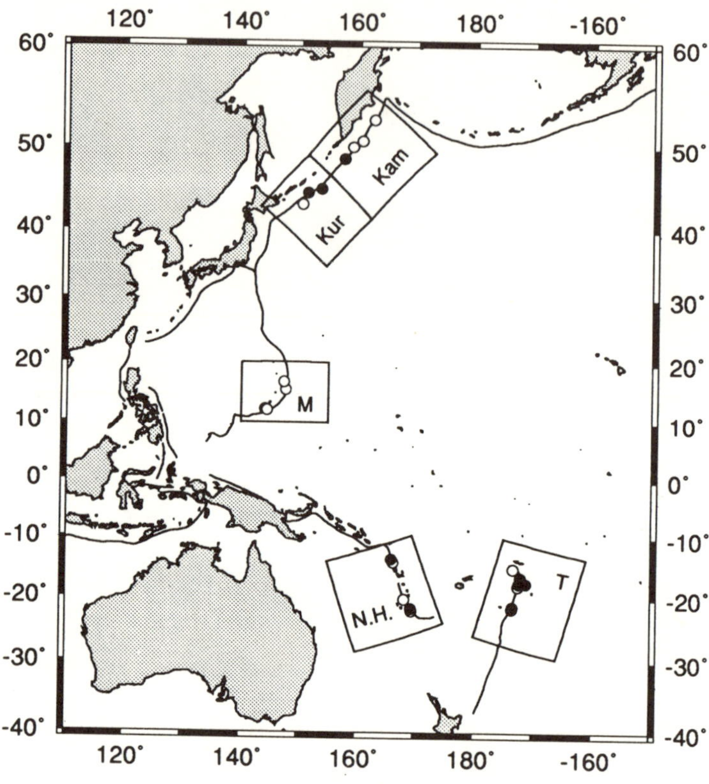

Figure 1

Locations of selected outer rise earthquakes (1980–1986) at the Tonga (T), New Hebrides (N.H.), Marianas (M), Kurile (Kur), and Kamchatka (Kam) subduction zones. Open circles represent tensional earthquakes, solid circles for compressional ones.

homogeneity by refining depths of all the events using the same body-wave modeling methodology.

In the depth determinations, vertical synthetic seismograms are constructed from the direct P, ocean bottom reflected $_pP$, $_sP$, and ocean surface reflected $_{pw}P$, $_{sw}P$ phases; transverse synthetic seismograms are constructed from the direct SH and ocean bottom reflected $_{SH}SH$ phases. For each earthquake, there are at least five stations with good recordings of vertical and horizontal ground displacements. The Preliminary Reference Earth Model (PREM, DZIEWONSKI and ANDERSON, 1981) with a 5 km average ocean depth is used in the waveform modeling. Since most of the selected outer rise earthquakes have magnitudes between 5.0 and 6.5, we assume the source time functions to be trapezoids with one second rise time for all depth determinations. Focal depth and duration of the source time function trade off in their contributions to synthetic waveforms. The best depth is determined from the post fit residual χ^2 as we perform a two parameter search for the depth and duration of the source time function simultaneously. The uncertainty range of an earthquake depth is chosen to be such that lower and upper limits of depths correspond to residuals 1.3 times the minimum residual. Figure 2 depicts an example of the fit of synthetics to data for the May 11, 1983 outer rise earthquake in the Tonga trench. Approximately the first minute of each teleseismic record is used to match the synthetics at each station. The residual is plotted in the upper right corner of Figure 2. In this example, the best depth is 35 km, measured from the surface of the plate. The duration of the source time function is 5 sec. We compare our refined depths for the selected outer rise earthquakes with the depths given by other studies documented by CHRISTENSEN and RUFF (1988) and find a maximum depth difference of 20 km. In Appendix 1 we show the fit of synthetic seismograms to observed seismograms for several outer rise earthquakes whose depths determined in this study are more than 10 km different from the values given by other studies.

Methodology

Calculation of Stresses

To approximate conditions in the oceanic lithosphere, we employ a two-dimensional, isotropic, layer over half-space, elastic earth model with constant gravity. More complicated models seem unnecessary since available earthquake data cannot completely constrain the parameters in complicated models. We use the static propagator to calculate displacements and stresses in a 2-D elastic bending plate floating on a "weak-fluid" half-space (WATTS *et al.*, 1980). Appendix 2 gives a detailed description of the formulation of the stress calculation.

The explicit time dependent behavior of the stresses depends on the viscoelastic composition of the lithosphere and mantle. Because of the highly underdetermined nature of the problem, we consider only a simple stress approximation. The viscoelastic relaxation effect is approximated by considering response of a relaxed earth. For the forces applied at the subducting slab for durations longer than the mantle relaxation time, a relaxed rigidity is used. We allow only shear stresses to relax. The Lamé constant λ remains the same as the seismic Lamé constant since it does not vary greatly in time (under the same confining pressure and other conditions, rock would not be significantly more compressed as time elapses). The modeling normal load, $f_z(x)$ (normal stresses on top of the entire plate) represents a synthetic force in the direction normal to the slab surface, which includes overburden pressure, slab pull component normal to the plate surface, and possibly induced mantle flow pressure. The modeling tangential load, $f_x(x)$ (shear stresses over a maximum width of 200 km at the interplate contact) simulates the mechanical coupling at the interplate contact and the effect of slab pull and the mantle

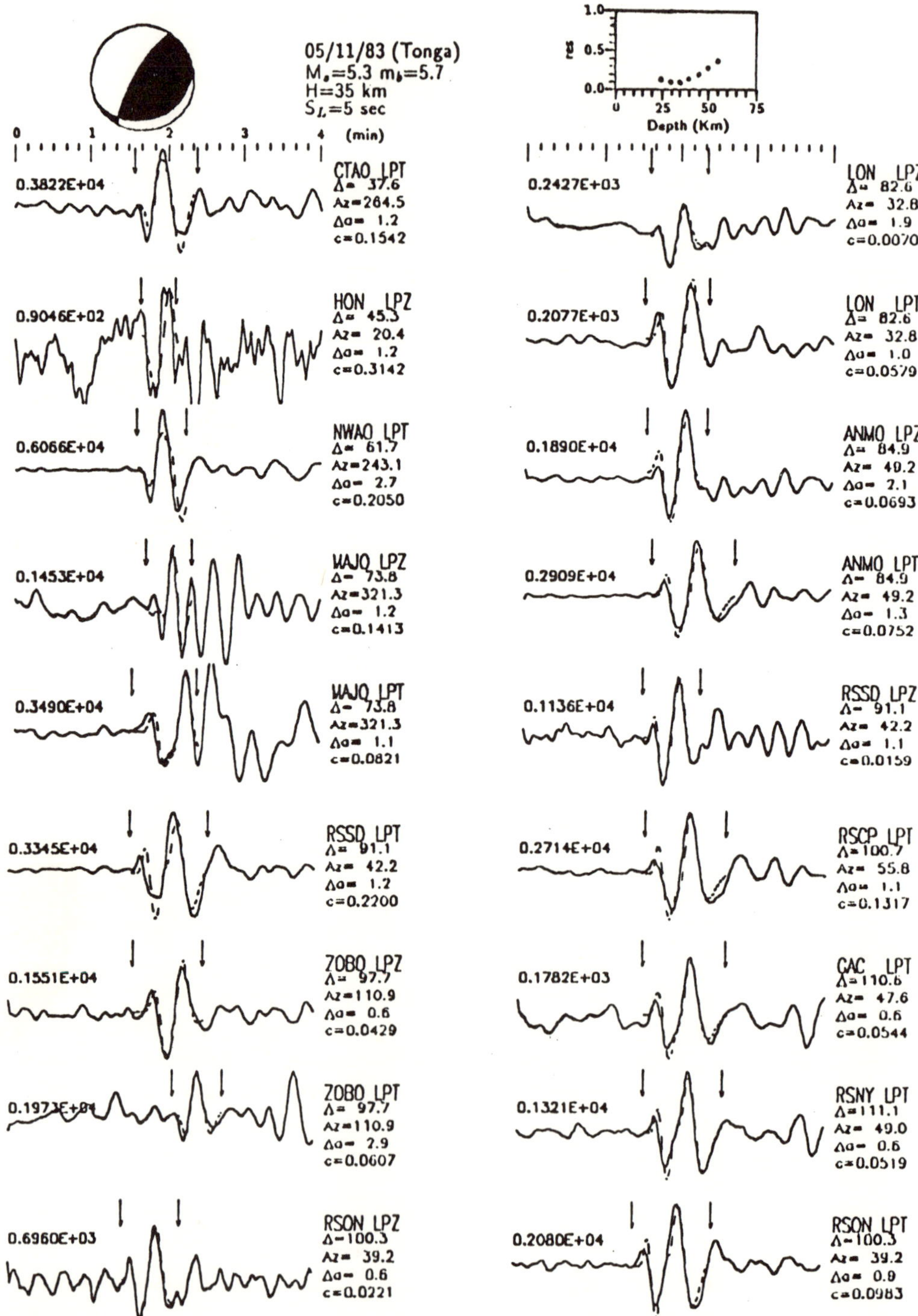
05/11/83 (Tonga)
M_s=5.3 m_b=5.7
H=35 km
S_L=5 sec

0 1 2 3 4 (min)

res
1.0
0.5
0.0
0 25 50 75
Depth (Km)

0.3822E+04
CTAO LPT
Δ= 37.6
Az=264.5
Δα= 1.2
c=0.1542

0.2427E+03
LON LPZ
Δ= 82.6
Az= 32.8
Δα= 1.9
c=0.0070

0.9046E+02
HON LPZ
Δ= 45.3
Az= 20.4
Δα= 1.2
c=0.3142

0.2077E+03
LON LPT
Δ= 82.6
Az= 32.8
Δα= 1.0
c=0.0579

0.6066E+04
NWAO LPT
Δ= 61.7
Az=243.1
Δα= 2.7
c=0.2050

0.1890E+04
ANMO LPZ
Δ= 84.9
Az= 49.2
Δα= 2.1
c=0.0693

0.1453E+04
WAJO LPZ
Δ= 73.8
Az=321.3
Δα= 1.2
c=0.1413

0.2909E+04
ANMO LPT
Δ= 84.9
Az= 49.2
Δα= 1.3
c=0.0752

0.3490E+04
WAJO LPT
Δ= 73.8
Az=321.3
Δα= 1.1
c=0.0821

0.1136E+04
RSSD LPZ
Δ= 91.1
Az= 42.2
Δα= 1.1
c=0.0159

0.3345E+04
RSSD LPT
Δ= 91.1
Az= 42.2
Δα= 1.2
c=0.2200

0.2714E+04
RSCP LPT
Δ=100.7
Az= 55.8
Δα= 1.1
c=0.1317

0.1551E+04
ZOBO LPZ
Δ= 97.7
Az=110.9
Δα= 0.6
c=0.0429

0.1782E+03
CAC LPT
Δ=110.6
Az= 47.6
Δα= 0.6
c=0.0544

0.1971E+04
ZOBO LPT
Δ= 97.7
Az=110.9
Δα= 2.9
c=0.0607

0.1321E+04
RSNY LPT
Δ=111.1
Az= 49.0
Δα= 0.6
c=0.0519

0.6960E+03
RSON LPZ
Δ=100.3
Az= 39.2
Δα= 0.6
c=0.0221

0.2080E+04
RSON LPT
Δ=100.3
Az= 39.2
Δα= 0.9
c=0.0983

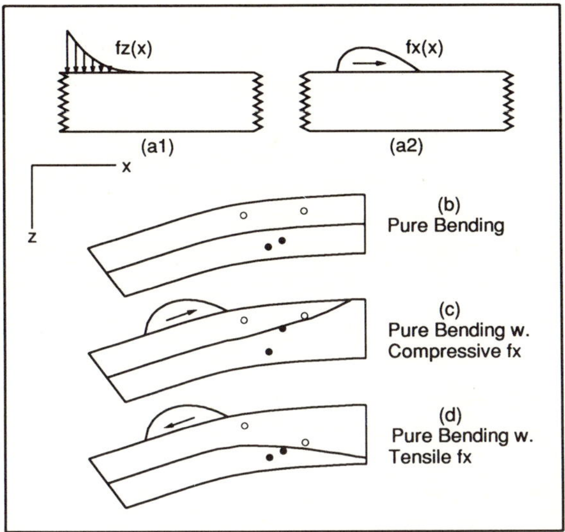

Figure 3

A cartoon of model for calculating stresses and estimating interplate coupling. x axis is horizontal, perpendicular to the trench, positive pointing to the ocean. z axis is positive downward. The stresses in response to normal load $f_z(x)$ in (a1) are calculated separately from the stresses resulting from tangential load $f_x(x)$ in (a2). The neutral surface (solid line) is in the central part of the plate in pure bending case, which satisfactorily explains the shallower tensional and deeper compressional outer rise events (b). With the occurrence of shallower compressional events (c), compressive $f_x(x)$ is applied at the plate contact to elevate the neutral surface to fit the data. Tensile $f_x(x)$ is needed to depress the neutral surface to fit the deeper tensional events (d).

shear exerted from the down-dip side of the slab. The effect of coupled width on the later inferred tectonic coupling will be elaborated in the discussion section. The elastic responses of a plate to the normal and tangential loads are added together to form the stresses in the bending plate (Figure 3(a1) and 3(a2)).

Estimation of Tangential Force (F_x)

The bending plate at an ocean trench is under horizontal deviatoric tension in the upper part and under deviatoric compression in the lower part. For the cases of pure bending (no tangential load), the neutral surface (zero stress boundary separating the tensional upper plate from the compressional lower plate) is located

---

Figure 2

Body-wave synthetics (dashed curves) and the observed seismograms (solid curves) for the May 11, 1983 outer rise earthquake in Tonga. On the right-hand side of each matching traces are printed the epicentral distance (Δ), Azimuth (Az), scaling factors (Δa and c). The residual vs. depth plot is shown in the upper right corner. For this earthquake, the refined depth is 35 km with 5 sec duration of the source time function.

at a depth of about half the plate thickness (Figure 3(b)). However, the depth of the neutral surface can be changed up-dip (toward and including the outer rise) if a tangential load is applied, and therefore, shallower compressional or deeper tensional outer rise earthquakes may be expected. We estimate the tangential load constrained by the best depths and the uncertainty ranges of our selected outer rise earthquakes at each subduction zone. A tangential load resisting plate subduction is required to explain shallower compressional earthquakes (Figure 3(c)), whereas a tangential load assisting plate subduction is needed to explain deeper tensional earthquakes (Figure 3(d)).

It is difficult to estimate the exact distribution of interplate coupling along the slab dip direction. Seismic moment release of interplate earthquakes can provide a rough picture of the coupling distribution in depth. A more strongly coupled base would be expected if shear resistance on the interface increases monotonically with lithostatic pressure up to the brittle-ductile transition that defines the base of the coupling zone (DAS and SCHOLZ, 1983). Stronger coupling at the down-dip edge of the coupling zone is also suggested by the study of down-dip underthrusting earthquakes at some segments of subduction zones (CHOY and DEWEY, 1988; SCHWARTZ et al., 1989). We define the envelope of tangential load magnitude as

$$f_x(x) = \tfrac{15}{4}\bar{f}_x\left(\frac{x}{w}\right)^{1/2}\left(1 - \frac{x}{w}\right), \quad 0 \leq x \leq w \tag{1}$$

where w is the width of coupling zone and $\bar{f}_x$ is the mean tangential load. This function describes the tangential load as reaching its maximum at about one-third of the coupling width measured from the down-dip edge, and decreasing to zero at the up-dip edge of the coupled zone.

Because tensional and compressional outer rise earthquakes are not always well separated in depth, a statistical approach is used to estimate the optimal neutral surface depth, and its corresponding tangential load. We introduce a penalty function

$$\Psi = \sum_{i=1}^{N} e^{\pm 10(d_i - h_i)/\sigma_i} \tag{2}$$

where h_i is the depth of the ith earthquake, σ_i is its depth uncertainty, d_i is the neutral surface depth right above or below (in depth) the ith event. The "+" sign is for compressional earthquakes and the "−" sign for tensional ones. The summation (2) is over the total number of earthquakes N at a given subduction zone. Thus Ψ is a measure of the goodness of fit of the neutral surface to all of the earthquakes for a given tangential load. The best fit of the neutral surface to the earthquake data corresponds to the minimum penalty function. Further, the optimal tangential load is associated with the best neutral surface depth. The selection of an exponential form in (2) is due to its faster increase in penalty value when the neutral surface depth violates the earthquake depths. Therefore, the search for an optimal tangential load is mainly constrained by those earthquakes whose depths are not consistent with the neutral surface depth for pure bending.

The introduction of the penalty function helps to find a best neutral surface and its corresponding tangential load given a set of earthquake depths. But by itself, the penalty function cannot provide uncertainty, an estimate of the tangential load from the uncertainty of the earthquake depths. We adopt a statistical procedure to search for the optimal tangential load and its uncertainty by allowing chatter in the earthquake depths used in (2). For each earthquake, penalty functions for a given stress model are evaluated at M depths, which spans a depth range approximately centered on the least waveform residual depth. For N earthquakes at a given trench, there are thus M^N combinations of the earthquake depths. A penalty function value Ψ_{jk} can be obtained by (2) for the jth tangential load and the kth combination of earthquake depths. For each M^N combination of earthquake depths, we find the best tangential load, which has the minimum penalty value. For each combination of earthquake depths, we also determine the probability for the optimal tangential load among the load models explored by

$$p_f = \prod_{i=1}^{N} P_i(F_{\zeta i}, N_P - n_\zeta, N_P - n_i) \tag{3}$$

where P_i is an integral probability of F value, $F_{\zeta i} = \chi_i^2 / \chi_\zeta^2$ (BEVINGTON, 1969, p. 196). N_P is the total number of points in the earthquake depth determinations, n_ζ and n_i are the numbers of modeling parameters in the depth determinations. Subscript ζ indicates the terms corresponding to the least residual depth, and subscript i, the ith depth designated by a given combination of earthquake depths. Finally, by weighting the best fitting tangential load with its probability p_f for each earthquake depth combination we can determine the mean and standard deviation of the best fitting tangential load at each subduction zone.

We introduce the tangential force (F_x) at the plate contact. F_x is a force per unit length along the strike of trench and is defined as

$$F_x = w\bar{f}_x \tag{4}$$

where w is the width of the coupling zone and $\bar{f}_x$ is the mean tangential load at the interplate contact. F_x is a measure of the sum of tangential forces applied on the subducting plate from the continental side. The larger the tangential load is, the narrower the coupling width is; or *vice versa*, if fitting the same earthquake data set. However, F_x does not vary greatly for different w and $\bar{f}_x$. Therefore, F_x can better characterize the coupling at each subduction zone. We use the coupling width as suggested by aftershock areas of great underthrusting earthquakes where available.

Estimation of Tectonic Coupling (F_{tc})

We should point out that although we have modeled F_x as a force acting at the interplate contact, this is an approximation. There are three major forces acting from the down-dip side of the plate (SPENCE, 1987): the tectonic coupling F_{tc}, the slab pull F_{sp}, and the mantle shear F_{sh} (Figure 4). F_{tc} is the total coupling force

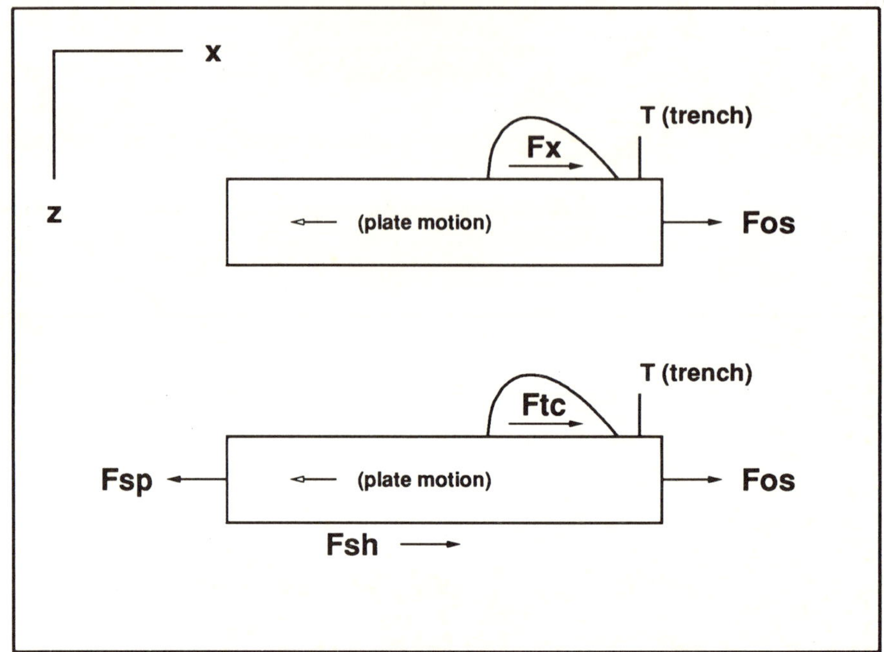

Figure 4

Force balance on a subducting slab (simplified in horizontal direction). The tangential load estimated from the outer rise events can be used to obtain the tangential force, F_x (upper panel), which is balanced by the total forces (F_{os}) from the ocean side. F_x is composed of interplate coupling force (F_{tc}), slab pull force (F_{sp}), and mantle shear force (F_{sh}) shown in the lower panel.

applied at the interplate coupling zone per unit length along strike. This force is resistant to relative motion of the plates. F_{sp} represents the total slab pull force on the entire slab, equivalently applied at the interplate coupling zone per unit length along strike. F_{sh} is the total mantle shear force on the slab, equivalently exerted at the interplate coupling zone per unit length along strike. To determine the direction of the mantle shear force, we assume that the mantle resists plate subduction. The simplified primary force equation at the down-dip side of a slab is

$$F_x = -F_{sp} + F_{tc} + F_{sh}. \tag{5}$$

Unlike F_{tc}, F_{sp} and F_{sh} are distributed along the entire length of the slab. The primary action of F_{sp} and F_{sh} is on the segment of the slab which is far (more than 100 km away) from the modeling region, the outer rise. Thus, the basic influence of the distributed F_{sp} and F_{sh} forces on the outer rise can be approximated as two single forces acting near the seismogenic zone, for our purpose we apply them at the interplate contact.

F_x is balanced by the force applied on the plate from the ocean side (F_{os}), which includes the ridge push and the mantle shear on the segment of the subducting slab before entering the trench. We need not specify the partitioning of these stresses. By rearranging (5), we can obtain

$$F_{tc} = F_x + F_{sp} - F_{sh}. \tag{6}$$

To estimate the slab pull F_{sp} and the mantle shear F_{sh} we first identify the mechanical neutral zone within each slab, and calculate the two forces associated with the portion of the slab above the neutral zone. All the forces acting on the slab downward from the neutral zone can be neglected since their net contribution to the part above the neutral zone is effectively zero. The neutral zone can be estimated from seismic studies, and we will elaborate on such an estimation in later sections.

The formula approximating F_{sp} is

$$F_{sp} = \Delta\rho g H L \sin(\alpha) \tag{7}$$

where $\Delta\rho$ is the average density difference between the subducting slab and the surrounding mantle due to the temperature contrast, given as $0.05 \, \text{g/cm}^3$ (YOKOKURA, 1981). $g = 10 \, \text{m/s}^2$ is the gravitational constant, H the slab thickness, L the length of the slab, and α the dipping angle of the subducting slab.

The mantle shear F_{sh} is estimated from the relation (DAVIES, 1980)

$$F_{sh} = (2L - w)\eta\dot{\varepsilon} \tag{8}$$

where w is the coupling width, L the slab length, η the mantle viscosity, and $\dot{\varepsilon}$ the mantle strain rate. We adopt the estimation of viscosity by CATHLES (1975): above 128 km depth, $\eta = 4 \times 10^{19}$ Pa s in a 75 km thick asthenosphere layer; below 128 km depth, $\eta = 10^{21}$ Pa s for a 600 km thick upper mantle. $\dot{\varepsilon} = v/l_0$, where v is the plate convergence rate, and l_0 the thickness of the viscous layers, taken as 75 km and 600 km above and below the 128 km depth, respectively.

Model Parameters

The plate thickness, density, Lamé constant, rigidity, normal load, and tangential load are the modeling parameters in our stress calculation. Given the thickness and material constants of the plate, the normal load can be inverted from the bending plate shape. This inversion for the normal load ensures that the calculated plate shape closely matches the actual plate shape. In this study the plate geometry comes from two kinds of data. Ocean bottom bathymetry data are used to specify the plate shape before they subduct into the trenches. The down-dip plate shapes are inferred from the subduction zone seismicity for each region (ISACKS and BARAZANGI, 1977). We choose densities of 3.2 and 3.3 g/cm^3 in the plate and the mantle, respectively. The lower value of the slab density than that for the mantle is required by the static equilibrium of our static stress model. Note that this density difference is not used to model the dynamic process of slab pull in plate subduction.

The Lamé constant λ does not vary much because the material of the earth is nearly incompressible. We use a typical seismological Lamé constant, 3×10^{10} N/m^2 in the plate.

The thickness and the rigidity of the plate are the two major parameters controlling the calculated plate stresses in our modeling. The calculated stresses are greater for larger rigidity (harder) and thicker plates, whereas they are less for smaller rigidity (softer) and thinner plates, for the same normal load. Here we use the elastic thickness of the plate given by the relationship of age to thickness, based on a depth and focal mechanism study of oceanic intraplate earthquakes by WIENS and STEIN (1983). A range of 37–64 km is used for elastic plate thickness for the subduction zones in this study. Plate ages are adopted from JARRARD (1986). For simplicity, we choose a one-layer plate model of uniform rigidity. The stress in the plate is largely associated with the long time response of the plate to the load. The rigidity of the plate used in our model should be less than the seismic rigidity, in response to the long time load of the mantle and the upper plate. The modeling rigidity is taken as 10^9 N/m^2, a value chosen subjectively to produce calculated stresses lying in a reasonable range. This rigidity is referred to as the relaxed rigidity, and is about an order of magnitude smaller than that commonly inferred from seismic data. We can fit the plate shape equally well for different rigidity values simply by changing the normal loads. Overall, this leads to uncertainty in absolute but not relative tangential loads on the plate. The calculated maximum deviatoric shear stress (τ_{max}) near the plate surface, is approximately 1–4 kb in the vicinity of the trench. This stress magnitude is compatible with the maximum bending stresses in the oceanic lithosphere estimated by WATTS et al. (1980). Their results indicate that the oceanic lithosphere near Oahu is capable of supporting stress of at least 1 kb from a yield stress envelope method in a rock mechanics study. We also compare the inverted normal load with the calculated overburden pressure from surface topography along a 200 km long cross section perpendicular to the Chile trench (which was also studied but is not discussed here). We find that the inverted normal load is about 20% larger than the calculated overburden pressure. This additional 20% of the normal load could be contributed by other factors, such as slab pull, mantle flow pressure and corner flow pressure. We infer that the magnitude of the inverted normal load in our simple stress modeling is on a reasonable order.

The material constants may vary within the plate, and a more realistic plate model must take into account such variations. Therefore we next examine the stress difference between two types of models, one with a uniform one-layer plate, the other with a multi-layer plate with varying rigidity and thickness, but with the same equivalent flexural rigidity (EFR) of the plate. The EFR is defined as $\sum \mu_i H_i$ where i is summed over all the layers in the plate. This definition of EFR is convenient for our test purpose and generally describes the plate flexure. Note that it is different from the conventionally defined flexural rigidity of the plate. We fix EFR to that

Table 2

Rigidities used in multi-layer plate models

Plate Thickness (km)	Rigidity (10^9 N/m^2)				H_m (km)	H (km)
	Layer #1 (0–20 km)	Layer #2 (20–40 km)	Layer #3 (40–60 km)	Layer #4 (60–80 km)		
40	0.714	1.286			26	20
60	0.370	0.667	0.963		38	30
80	0.227	0.409	0.591	0.773	52	40

H_m: Depth of neutral surface in multi-layer models
H: Depth of neutral surface in one-layer models

given by a one-layer plate model of 40 km thickness and uniform rigidity of 10^9 N/m^2. We construct three different layered plate models as indicated in Table 2. They generally have increasing rigidity with depth and plate thickness ranging from 40 to 80 km (the increase of rigidity with depth is taken from the PREM Earth model). For a given thickness of the plate, the neutral surface depths for the multi-layer plate models decrease by about 6–12 km with respect to the neutral surface depth in a plate of uniform rigidity. The depth variation of the neutral surface from the one-layer plate model to a multi-layer plate model is within the uncertainty range of the depths of outer rise earthquakes (5–15 km).

We also examine the influence of the depth dependent rigidity models on the estimation of tangential load constrained by the earthquake data in the Kamchatka region. Approximately 350 bar tangential load is needed to fit the depth of the shallowest compressional event in a uniform rigidity plate model; however, it requires 300 bar to fit that earthquake in a multi-layer plate model. The relative difference is about 14%.

From examining the effect of the rigidity model on stresses in the plate, we conclude that a one-layer plate model with uniform rigidity can give a relatively satisfactory approximation of the stresses in the plate, given the limited information of stresses at depth from outer rise earthquakes. 10–20% variations in stress magnitudes are associated with uncertainty in the rigidity model. Of course, this is within the context of our limited rheology, involving an effective viscoelasticity. Much greater stress uncertainties are associated with the possibility of alternate rheologies.

We are aware of the limitation of using an elastic plate model in our study. Some anelastic effects, such as plasticity and nonlinear creep are not optimally represented in such models. However, we utilize a simple model, an elastic one, assuming that tectonic stress is largely caused by elasticity in the upper 100 km of the lithosphere. The viscous relaxation effect of a bending plate is approximated by a largely reduced rigidity in the elastic plate. Viscous effects may be important in the bending process of subducting plates, but the lack of well-constrained material parameters in visco-elastic models makes it difficult to obtain results with improved certainty. ZHANG *et al.* (1985) explored the many trade-offs in alternate viscous parameterizations, finding that elasticity and plasticity are important, but not uniquely defined. Our models are for a particularly simple, plausible case, but clearly different conclusions may result if other rheologies are considered.

Results

We estimate the mean tangential loads ($\bar{f}_x$) and their standard deviations at the interplate contact at the Tonga, New Hebrides, Kurile, and Kamchatka subduction zones. The outer rise earthquake locations and modeled plate surface in a cross section perpendicular to the trench at each subduction zone are shown in Figure 5. The calculated deviatoric maximum shear stress τ_{max} in the plate near the trenches is between 1 and 4 kb for all the regions. The steep dip of the New Hebrides plate causes larger bending stress, as high as 4 kb.

At the Tonga subduction zone, there is a lack of great shallow underthrusting earthquakes during this century, thus no well-constrained interplate coupling width is available. The only documented interplate earthquake (12/19/82, $M_w = 7.5$) occurred near the intersection of the Louisville Ridge with the trench at the southern end of the Tonga region. This event is associated with the enhanced interplate coupling due to the buoyant Louisville Ridge subduction (CHRISTENSEN and LAY, 1988). However, lack of characteristic large or great interplate earthquakes during this century in Tonga does not necessarily rule out the capability of this region to produce large size interplate earthquakes. We assume the coupling width is 200 km and obtain $\bar{f}_x$ of 450 bar with 283 bar standard deviation. Note that different widths for coupled zones will not change the estimate of the interplate coupling (to be derived later) because the larger value of the width corresponds to a lower tangential load, or *vice versa*. The probability distribution for the optimal tangential load $\bar{f}_x$, $P(\bar{f}_x) = \sum_f p_f(\bar{f}_x)$ is shown in a histogram in Figure 6(a), with ξ and σ as mean and one standard deviation of $\bar{f}_x$ for all the combinations of earthquake depths. A positive tangential load indicates compressional shear stress at the interplate contact. The irregular histogram reveals the important roles some earthquakes play when their depths are close to the neutral surface depth.

Figure 5(a)

Selected outer rise earthquakes in map and cross section views at the Tonga, New Hebrides, and Marianas subduction zones. The background maps and aftershock areas of the great subduction zone earthquakes are adopted from CHRISTENSEN and RUFF (1988). Outer rise events are labeled in chronological order in each region. The down-dip plate geometry at each cross section is approximated by ISACKS and BARAZANGI (1977) based on the projection of the depths of the down-dip seismicity (1964–1989) in the box onto the vertical plane A-A'. We also plot the PDE locations of down-dip seismicity between 1964 and 1989 in each cross section.

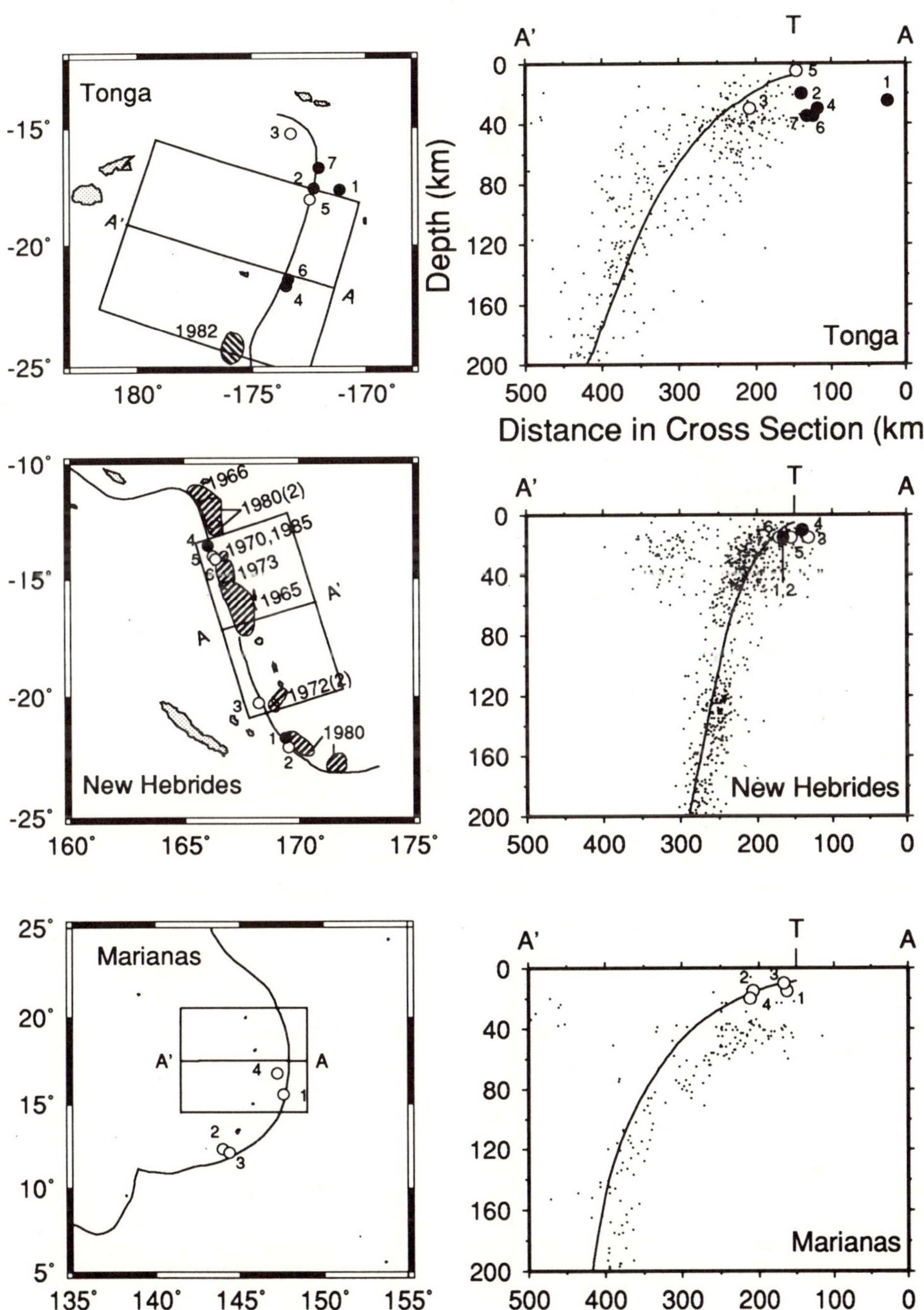
Tonga
-15°
-20°
-25°
180° -175° -170°
3
7
2 1
5
6
4
1982

A' T A
0
40
80
120
160
200
Depth (km)
500 400 300 200 100 0
Distance in Cross Section (km)
5
3
2 1
4
7 6
Tonga

New Hebrides
-10°
-15°
-20°
-25°
160° 165° 170° 175°
1966
1980(2)
4 1970,1985
5
6 1973
1965 A'
A
1972(2)
3
1 1980
2

A' T A
0
40
80
120
160
200
500 400 300 200 100 0
4
6 3
5 8
1,2
New Hebrides

Marianas
25°
20°
15°
10°
5°
135° 140° 145° 150° 155°
A' 4 A
1
2
3

A' T A
0
40
80
120
160
200
500 400 300 200 100 0
2 3
1
4
Marianas

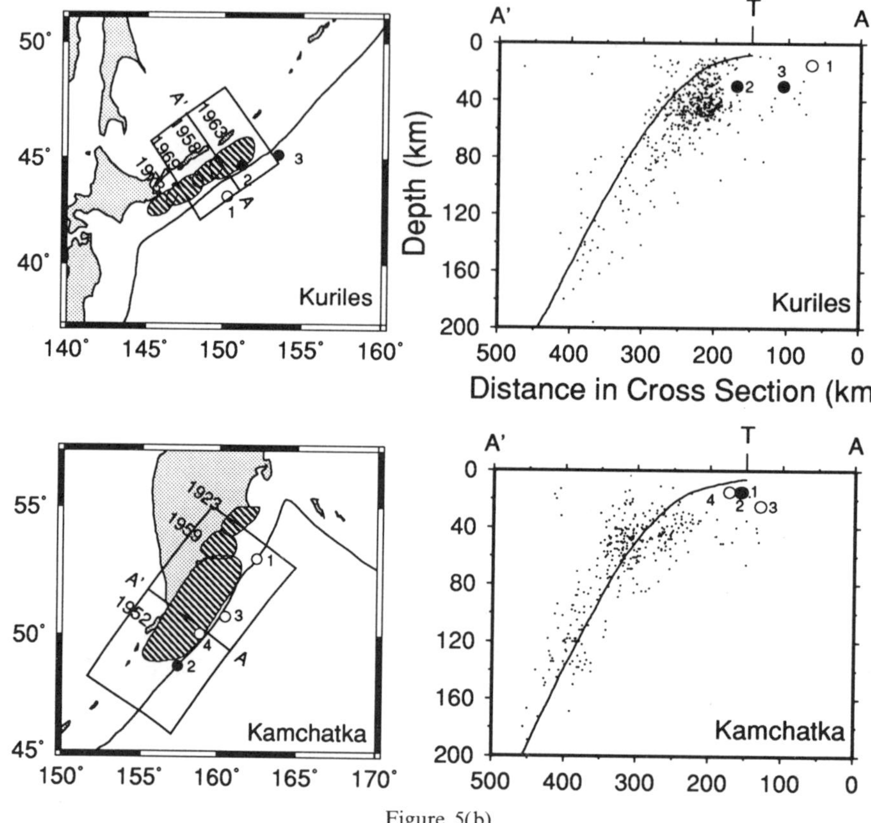

Figure 5(b)

Selected outer rise earthquakes in map and cross-section views at the Kurile and Kamchatka subduction zones. Other details as in Figure 5(a).

In Figure 6(b), we show the calculated deviatoric maximum shear stress τ_{max} in the plate under the statistical mean $\bar{f}_x$ (obtained from (a)) along with the earthquake depths and their uncertainties. The solid curve inside the plate is the neutral surface of the calculated principal stresses under the mean $\bar{f}_x$. The two dashed

Figure 6

Search for the optimal $\bar{f}_x$ and its uncertainty using penalty function at the Tonga subduction zone. The histogram in (a) depicts the frequency $P(\bar{f}_x)$ of the searched best $\bar{f}_x$ for all the depth combinations, along with the mean (ξ) and one standard deviation (σ) of $\bar{f}_x$. Positive $\bar{f}_x$ stands for compressive load. The calculated maximum deviatoric shear stress τ_{max}, denoted by crosses (the length of a cross indicates the stress magnitude) under ξ is shown in (b) with our outer rise events. The solid curve inside the plate is the neutral surface under ξ. The two dashed lines above and below the solid curve correspond to the neutral surfaces given by tangential loads $\xi \pm \sigma$. Shown in the upper right corner are the side view back projections of focal mechanisms of our outer rise events when viewing from SW to NE direction at an angle of N23°E along the trench strike. Estimated best $\bar{f}_x$ and the associated neutral surface for the enlarged data set (the solid curve) is shown in (c), as compared with the neutral surface given by ξ for our data set (the dashed curve). Additional events in the enlarged data set are listed in Table 1b.

curves above and below the neutral surface correspond to the neutral surfaces within one standard deviation of the mean $\bar{f}_x$. To give more information about the outer rise earthquakes used to constrain the $\bar{f}_x$, side view (horizontal view along plate strike) focal mechanisms of the earthquakes are plotted in their chronological order, along with body-wave magnitudes on the right side of Figure 6(a). Figure 6(c) shows results from an enlarged data set which we discuss later.

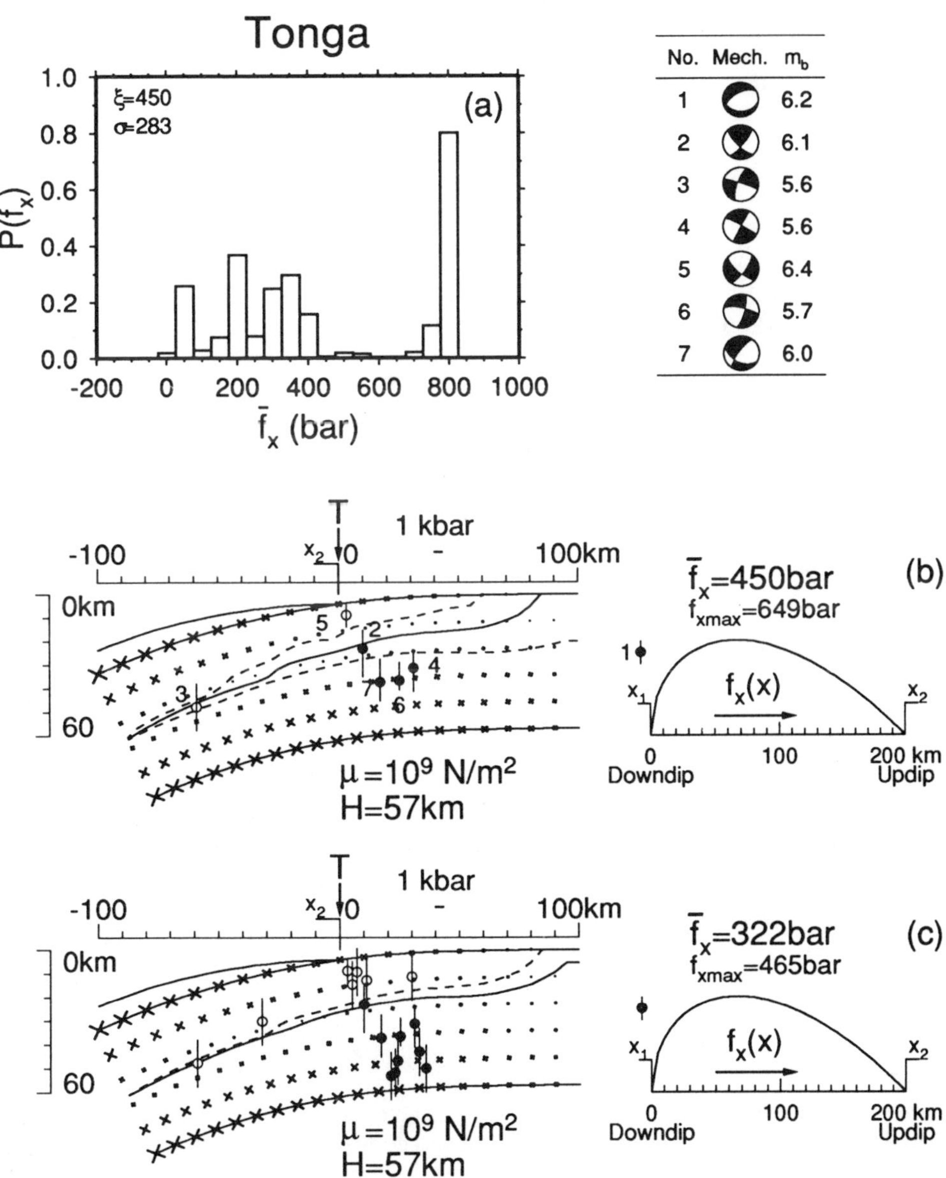

The outer rise earthquakes associated with the New Hebrides trench are located near the northern and southern segments of the subduction zone. The depths of the tensional earthquakes overlap those of the compressional events. We obtain a compressive 671 ± 186 bar $\bar{f}_x$ (Figure 7). We assume the coupling width to be 125 km, as suggested by the 1965 interplate earthquake which has the largest measured rupture area in the New Hebrides.

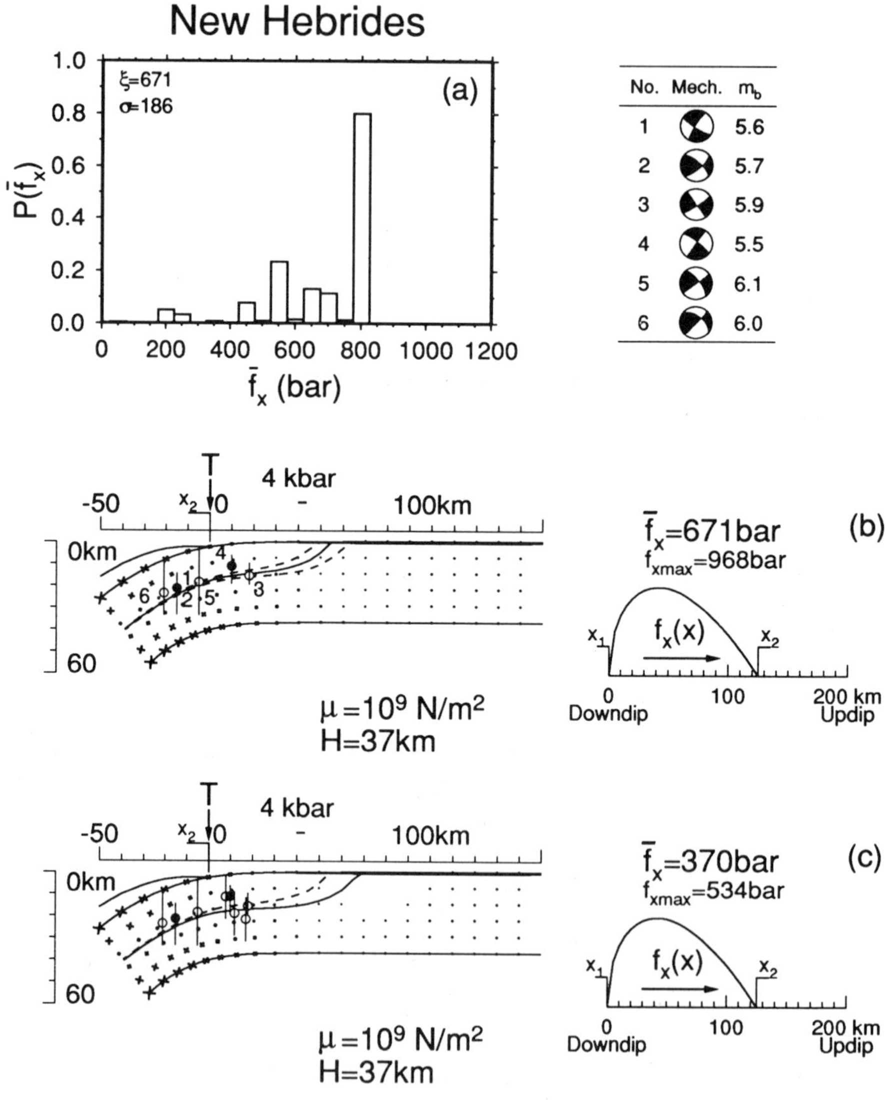

Figure 7

Search for the optimal $\bar{f}_x$ and its uncertainty using penalty function at the New Hebrides subduction zone. The side view back projections of focal mechanisms of the outer rise events are made on a plane normal to the N24°W, from SE to NW direction along the trench strike. Other details as in Figure 6.

For the Kurile subduction zone, a 166 ± 85 bar compressive $\bar{f}_x$ is obtained (Figure 8). We assume a coupling width of 130 km, suggested by the aftershock area of the 1963 great earthquake.

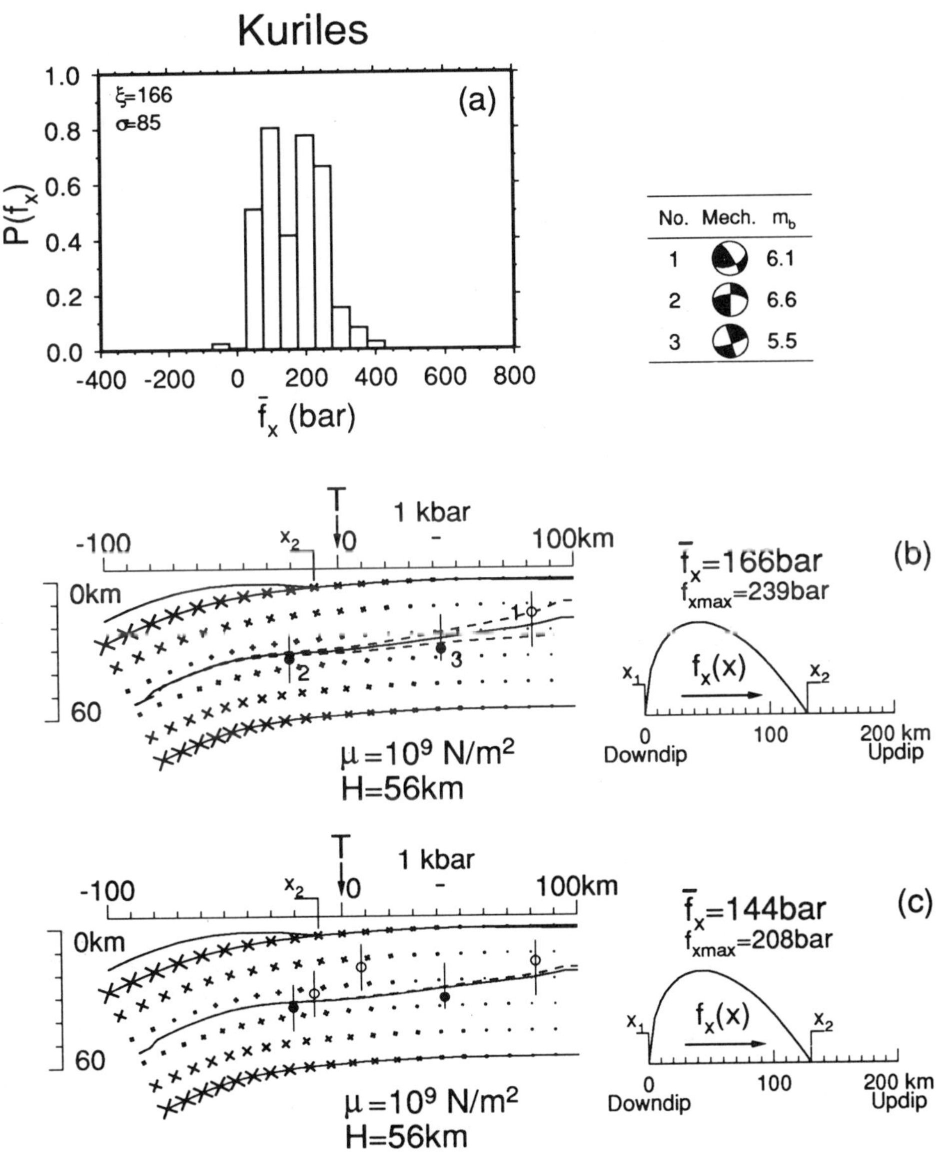

Figure 8

Search for the optimal $\bar{f}_x$ and its uncertainty using penalty function at the Kurile subduction zone. The side view back projections of focal mechanisms of the outer rise events are made on a plane normal to the N12°E, from SW to NE direction along the trench strike. Other details as in Figure 6.

For the Kamchatka subduction zone, the compressive $\bar{f}_x$ is estimated to be 266 ± 49 bar (Figure 9). We use a coupling width of 200 km, based on the 1952 great underthrusting earthquake.

Several tensional outer rise events have been identified at the Marianas subduction zone, but no compressional events were found. These tensional events occurred in the upper part of the plate, consistent with a pure bending model with no tangential load (Figure 10). But zero tangential load is not a unique solution, given the absence of compressional or deep tensional events in our data set. Lack of observed compressional outer rise earthquakes may reflect nearly total decoupling in the region, consistent with the absence of large interplate thrust earthquakes.

The estimated $\bar{f}_x$ and its standard deviation at each subduction zone is listed in Table 3. In order to further test the stability of these results, we apply the stress modeling to enlarged data sets for the Tonga, New Hebrides, and Kurile subduction zones. The enlarged data sets include our earthquake data and data compiled by CHRISTENSEN and RUFF (1988). CHRISTENSEN and RUFF (hereafter CR)'s data exhibit more scatter in depth possibly because they are compiled from different

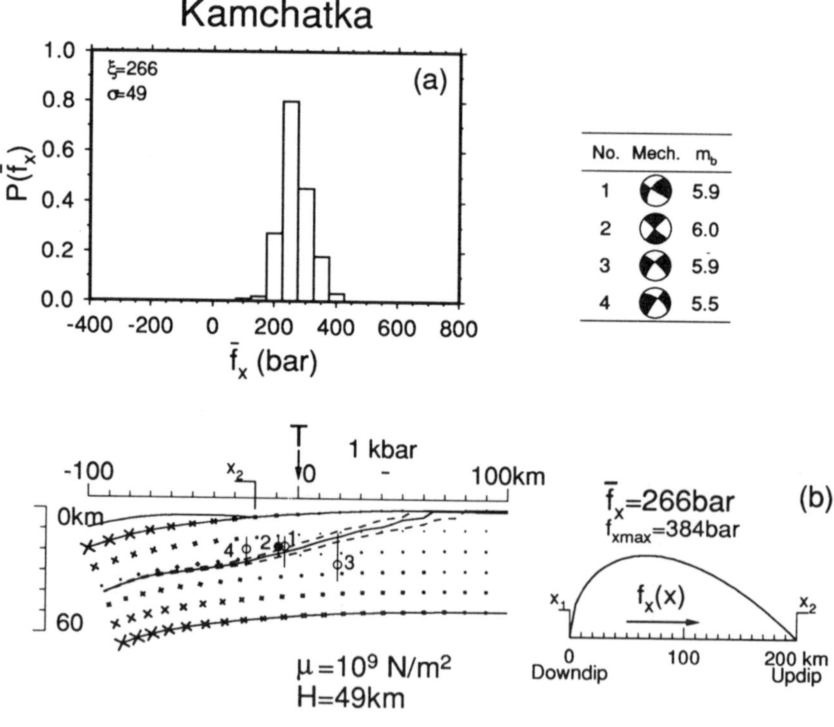

Figure 9
Search for the optimal $\bar{f}_x$ and its uncertainty using penalty function at the Kamchatka subduction zone. The side view back projections of focal mechanisms of the outer rise events are made on a plane normal to the N12°E, from SW to NE direction along the trench strike. Other details as in Figure 6.

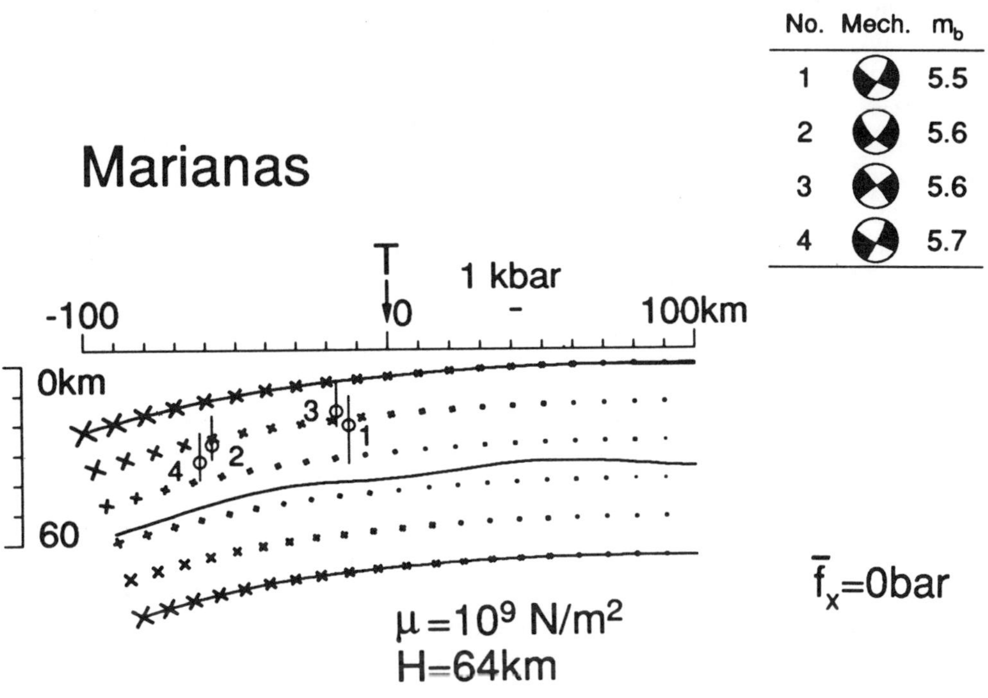

Figure 10

Calculated τ_{max} under pure bending condition ($\bar{f}_x = 0$) at the Marianas. The side view back projections of outer rise earthquakes are made on a plane normal to the local strike direction along the trench. Other details as in Figure 6. The depths of tensional outer rise earthquakes can be well explained by the pure bending neutral surface.

Table 3

Estimated mean tangential load $\bar{f}_x$ and tangential force F_x

Trench	Plate Age (Ma)	Plate Thickness (km)	Coupling Width (km)	$\bar{f}_x$ (bar)	F_x (10^{12} N/m)
New Hebrides	52	37	125	671 ± 186	8.39 ± 2.32
Kamchatka	90	49	200	266 ± 49	5.32 ± 0.98
Kurile	119	56	130	166 ± 85	2.16 ± 1.11
Tonga	120	57	200	450 ± 283	9.00 ± 5.66
Marianas	155	64	N/A	0	0

sources and derived by different methods. But their data span a longer time window (1964–1986), and therefore involve a larger population. When combining our data set with CR's, we take our earthquake depth solution rather than theirs for common events. We also retain our earthquake depth uncertainties. Only those events in CR's data set with depths determined by individual studies other than

PDE and HCMT, are used to construct the combined data. We assume $\pm 10\,\mathrm{km}$ uncertainty for CR's data. There are no qualified events in CR's data set for the Kamchatka subduction zone, thus no combined data set is available for estimating $\bar{f}_x$ in this region. CR's data set does not contain any compressional events in the Marianas subduction zone, so no combined data is available in the Marianas. The outer rise earthquakes from CR selected to construct the combined data, are listed in Table 1b. The estimated neutral surfaces for the enlarged data sets are shown by solid curves in the third panel (c) in each of Figures 6–8. The dashed curves show the preferred result for our data alone. Comparing $\bar{f}_x$ estimated from the combined data set with the estimates from our data set, we find no drastic changes of the calculated neutral surface depth. The differences in the neutral surfaces are usually less than 5 km, with most differences being in the offshore extrapolation. The magnitudes of $\bar{f}_x$ vary by as much as a factor of two, as for the New Hebrides, but this is due to large scatter in the expanded data set. The results are very similar for the other regions.

Using the estimated $\bar{f}_x$ and the assumed coupling width at each subduction zone in (4), we calculate the tangential forces F_x at each subduction zone (Table 3). F_x ranges from 10^{12} to $10^{13}\,\mathrm{N/m}$ and has the error bands shown in Figure 11. The estimated $\bar{f}_x$ from the enlarged data sets are also converted into F_x, and are marked by a shaded vertical bar for each region (Figure 11). The estimated $\bar{f}_x$ from the combined data are within one standard deviation of those from our data for the

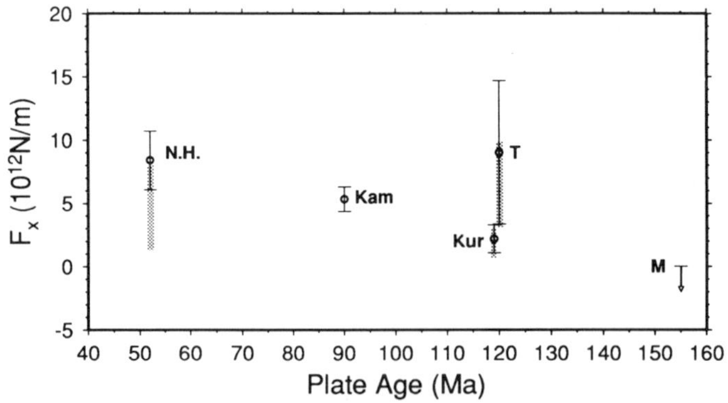

Figure 11

Estimates of F_x inferred from our outer rise earthquakes (denoted by solid dots with error bars) versus plate age at the Tonga (T), New Hebrides (N.H.), Kurile (Kur), Kamchatka (Kam), and Marianas (M) subduction zones. The vertical shaded bars represent one standard deviation range of F_x derived from the enlarged data sets. The discrepancy in F_x based on the enlarged data from that based on our data is within one standard deviation for Tonga and the Kuriles, and slightly more than one standard deviation of our $\bar{f}_x$ for the New Hebrides. No additional events are available to construct the combined data set for Kamchatka, thus no corresponding estimate of F_x is made there. No estimate of F_x is made for the combined data set for the Marianas also since earthquake data there cannot constrain a close-end uncertainty in $\bar{f}_x$.

Tonga and Kurile subduction zones. The estimated $\bar{f}_x$ from the combined data for the New Hebrides falls within about two standard deviations of the $\bar{f}_x$ from our data due to the presence of some deeper tensional events in CR's data set. We feel the estimated $\bar{f}_x$ from our data is generally stable, within one standard deviation, for characterizing the interplate coupling.

We also calculate the slab pull F_{sp} and mantle shear F_{sh} at each subduction zone according to (7) and (8) (results listed in Table 4). The coupling width for the Marianas subduction zone is assumed to be 100 km. The tectonic coupling F_{tc} is then obtained from (6) and is shown in Figure 12. The slab pull correction requires information regarding the effective length of the deep slab, the slab thickness, and the mean density difference between the slab and the surrounding mantle. For the Kamchatka, Kurile, Tonga and Marianas arcs, we estimate the effective lengths of the slabs (Table 4) by averaging the position of the transition from down-dip tension to down-dip compression neutral seismic zone at each region delineated by ZHOU (1990) from focal mechanism studies of intermediate and deep earthquakes. For the New Hebrides, seismicity in the slab shows a preponderance of downdip tensional events between 100–200 km in depth (LAY *et al.*, 1989), implying that they are induced by slab pull. BURBACH and FROHLICH (1989) found the intermediate-depth seismicity extends to depths of 200–300 km with a dip of 65–70°. We estimate the effective slab depth to be 250 km, based on the assumption that the seismic quiescence below 250 km depth reflects neutral or weak tectonic stresses or, more likely, the absence of slab. The corresponding effective length of the slab is assumed to be approximately 270 km for the New Hebrides. We use the same slab thicknesses as in the stress modeling, which corresponds to the 750°C isotherm. A density difference of 0.05 g/cm^3 is adopted from YOKOKURA (1981), based on numerical results of HASEBE *et al.* (1970) and TOKSÖZ *et al.* (1971, 1973).

Basically we have estimated two force terms at five subduction regions, F_x as the total tangential force acting on the slab from the subduction side, and F_{tc} as the shear force exerted over the interplate coupled zone. Lower F_{tc} have been found for Kamchatka and the Kuriles, around 6×10^{12} N/m; and higher F_{tc}, around

Table 4

Calculated F_{sp}, F_{sh}, and F_{tc}

Trench	Slab Length (km)	Coupling Width (km)	Rate (cm/yr)	Dip Angle (°)	F_{sp} (10^{12} N/m)	F_{sh} (10^{12} N/m)	F_{tc} (10^{12} N/m)
New Hebrides	270	125	2.7	44	3.47	0.35	11.50 ± 2.32
Kamchatka	150	200	9.3	25	1.55	0.16	6.72 ± 0.98
Kurile	500	130	9.3	28	6.57	2.90	5.84 ± 1.11
Tonga	250	200	8.9	28	3.34	0.45	11.90 ± 5.66
Marianas	200	100	4.0	24	2.60	0.20	<2.40

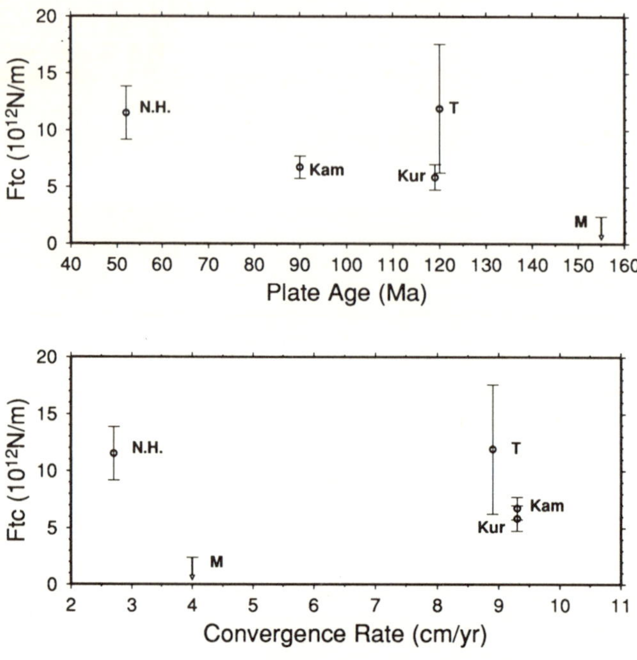

Figure 12

Estimates of F_{tc} inferred from outer rise earthquakes in the Tonga (T), New Hebrides (N.H.), Kurile (Kur), Kamchatka (Kam), and Marianas (M) subduction zones. Upper panel depicts the F_{tc} versus plate age, and lower panel, versus plate convergence rate. The F_{tc} for Tonga and the New Hebrides are about 2 times greater than that for the Kuriles and Kamchatka. No significant correlations of F_{tc} with either age or convergence rate can be seen from our data.

12×10^{12} N/m for Tonga and the New Hebrides. We can only estimate the upper limit of F_{tc} at the Marianas as 2.4×10^{12} N/m, and cannot preclude a value of zero.

Discussion

Shear stresses at subduction zones have been estimated by a variety of means. We now focus on the magnitude of the stresses in these estimates. MOLNAR and ENGLAND (1990) modeled the heat generation and heat transfer at the interplate contact and gave a range of 300–1000 bar for the shear stress, based on heat flow data from the Japan and Peru subduction zones. VAN DEN BUEKEL and WORTEL (1988) used a model of temperature and pressure dependent rheology to obtain an average shear stress of 100–400 bar at the plate interface constrained by the heat flow data between the trench and volcanic line at the Japan subduction zone.

We obtain $\bar{f}_x$ at four coupled subduction zones: Kamchatka, Kurile, Tonga and New Hebrides (the Marianas is considered an uncoupled subduction zone). The $\bar{f}_x$

(i.e., shear stress at the interplate contact) range from 166–671 bar and appear to be compatible with the estimated shear stresses given by MOLNAR and ENGLAND (1990) and VAN DEN BUEKEL and WORTEL (1988).

Tectonic Coupling F_{tc}

We find that the estimated $\bar{f}_x$, with one standard deviation, can explain all of our selected outer rise earthquake depths within their uncertainties at each subduction zone. Thus the associated F_{tc} may represent the tectonic coupling that is present in our chosen time window, for each subduction zone; if F_{tc} has changed during the time window, the variation is likely to be within the one standard deviation interval.

The tectonic coupling (F_{tc}) in Tonga and the New Hebrides, although subject to large uncertainties, appears higher than F_{tc} at the Kuriles and Kamchatka. This does not seem consistent with the observation that Tonga and the New Hebrides are probably less seismically coupled than the Kuriles and Kamchatka. How can we explain this result?

If Tonga and the New Hebrides indeed have greater tectonic coupling, one of the causes could be the material properties at the plate contact. In Tonga and the New Hebrides we have the case of an oceanic plate subducting under an oceanic plate, while for the Kurile and Kamchatka regions the overriding plate is continental. An oceanic plate generally is more homogeneous and less fractured, hence stronger than a continental upper plate. Rock mechanics experiments reveal that basalt has higher strength than quartzite (JAEGER, 1979), suggesting that oceanic lithosphere is stronger than continental lithosphere due to different materials. CHEN and MOLNAR (1983) analyzed global intracontinental and intraplate earthquakes and their thermal-mechanical implications, suggesting that oceanic lithosphere is much stronger than continental lithosphere. Therefore ocean-ocean plate contact may sustain larger stress, thus, higher tectonic coupling.

On the other hand, the estimates of F_x at Tonga and the New Hebrides may be biased by laterally-heterogeneous fault strength. The subduction of bathymetric features on the oceanic plate may cause heterogeneous fault strength at the plate interface (e.g., KELLEHER *et al.*, 1974; RUFF, 1989). It is likely that a fault with heterogeneous strength may correspond to overall weaker seismic coupling than one with homogeneous strength; however, the localized maxima in strength can be greater than an overall average. The higher estimates of tectonic coupling at Tonga and the New Hebrides may be associated with locally heterogeneous fault strengths there, since bathymetric features are subducting near the northern corner and near central Tonga and near the northern and central New Hebrides trenches (ZHANG and LAY, 1992). These are regions where our outer rise earthquakes happen to be located. Local variations may also help to explain the overlap of tensional and compressional focal mechanisms at depth for Tonga and the New Hebrides and the resulting large uncertainties for the estimated F_x. In contrast, there are no major bathymetric features being subducted along the segments of the Kurile and Kamchatka trenches where our outer rise earthquakes are located. Any attempt to model full three-dimensional stress fields is beyond our computational and data resources at present.

RUFF (1989) found a correlation between the excess trench sediment and great interplate thrust earthquakes at some subduction zones. He suggested that the excess trench sediment implied subduction of a coherent sedimentary layer, which will create a smooth seismic strength at the interplate contact, thus causing great interplate earthquakes. He also pointed out that sediment subduction may not be the sole cause of great earthquakes because there are some regions of no excess trench sediments which have great interplate earthquakes, or some regions of excess trench sediments which have no great interplate earthquakes. There may be effects of sediment subduction on fault strength. However, estimates of F_{tc} in our study cannot provide information for examining this speculation. The sediment thickness for Tonga, the New Hebrides, and the Kuriles are very small and about the same, 0.4 km; the sediment thickness is not available for Kamchatka (SCHOLL, 1990). What we might speculate is that the thin sediment layers being subducted here may limit the available water to elevate pore fluid pressure and reduce fault strengths.

Back-arc spreading processes at subduction zones tend to reduce the normal force between the subducting plate and the upper plate (e.g., UYEDA and KANAMORI, 1979). This will result in lower tectonic coupling in the regions with active back-arc spreading. We find the values of F_{tc} in our study are not positively correlated with the presence of back-arc spreading, suggesting that back-arc spreading may be a secondary effect on tectonic coupling since it is a passive process.

The higher values of F_{tc} in Tonga and the New Hebrides are mainly caused by greater background stress due to severe plate bending and the distribution of the outer rise earthquakes towards the inland direction. We examine the influence of uncertainties in the plate shape and locations of outer rise earthquakes on the estimates of F_{tc} for Tonga by varying plate dip angle by $10°$ and shifting outer rise earthquake locations by 20 km towards the offshore direction. We find that the distribution of outer rise earthquakes is relatively more important in controlling the estimates of F_{tc} than the plate shape and that the absolute value of the estimates of F_{tc} can be affected by as much as 40%.

The oceanic plate can be fractured by faults, which will probably form at shallow depths. The overall shallow strength of the oceanic plate is, thus, weakened. The approximation of a uniformly elastic plate to such problem may be inaccurate. To make a qualitative guess about how intraplate faults affect our estimated F_{tc}, we can set a lower value of rigidity in the top layer of the plate. This will result in stress values smaller with respect to those in a uniform elastic plate, and the shift of pure bending neutral surface to deeper depth. The estimated F_{tc} might remain the same for the same data set, however, because the smaller background stress effect on F_{tc} might be canceled by that of the deeper neutral surface.

The influence of the widths of the coupled zones employed in our study is very small because F_{tc} depends on the product of the coupled width and the tangential load. Tangential load trades off with the coupled width in order to fit the same data set. The coupled width may be overestimated because random errors in the aftershock locations will result in enlarged areas.

The uncertainties for F_{sp} and F_{sh} are mainly associated with how well we know the density contrast between the slab and the surrounding mantle ($\Delta\rho$), the effective length of the slab (L), the coupled width (w), the slab thickness (H), the slab dip angle (α), and the mantle viscosity (η). We take typical values for $\Delta\rho$ of 0.05 g/cm^3, L of 500 km, w of 150 km, H of 40 km, α of 30°, and η of 10^{21} Pa s. If we assume 100% uncertainty for the mantle viscosity and 20% uncertainty for the remaining parameters mentioned above, we will obtain 2×10^{12} N/m and 5×10^{12} N/m for the uncertainty of F_{sp} and F_{sh}, respectively. The uncertainty for F_{sp} is about the same as itself; whereas the uncertainty for F_{sh} is about one order of magnitude greater than itself. These uncertainties will perturb our estimates of F_{tc} by about 50% for the Kuriles and Kamchatka, and 25% for Tonga and the New Hebrides. Of course, this is only an *ad hoc* estimate for the uncertainties of the material constants, not including the model uncertainties. As it is well known, a rigorous estimate of uncertainty is almost impossible at present because of so many assumptions for models and uncertainties for the parameters used in such studies.

Influence of Temporal Variations in F_{tc} on Outer Rise and Down-dip Subduction Seismicity

From another independent source, we estimate the earthquake coupling force F_{ec} at each subduction zone (listed in Table 5). The estimates are based on the moment magnitude (RUFF and KANAMORI, 1980, 1983), length (L), and width (W) of the rupture zone of the largest earthquakes for each subduction zone. The

Table 5

Estimated earthquake coupling force F_{ec}

Trench	Characteristic Earthquake	M_w	Rupture Width (km)	Rupture Length (km)	F_{ec} (10^{12} N/m)
New Hebrides	09/20/20	7.9	100	143	0.14
Kamchatka	01/04/52	9.0	200	650	0.69
Kurile	10/13/63	8.5	150	250	0.43
Tonga	04/30/19	8.3	160	228	0.22
Marianas	1929, 1931	7.2	N/A	N/A	N/A

moment magnitude and stress drop are given by

$$M_w = \tfrac{2}{3}(\log_{10} M_0 - 16.1) \qquad (9)$$

$$\Delta\tau = 2\mu\,\Delta\varepsilon = 2\mu\,\frac{D}{W} \qquad (10)$$

where μ is seismic rigidity of $10^{10}\,\text{N/m}^2$, ε the strain, and D the coseismic displacement on a fault plane. Assuming the rupture area is an ellipsoid with $W/2$ and $L/2$ as the shorter and longer axes, the seismic moment release can be expressed as

$$M_0 = \mu D A = \frac{\pi\mu}{4}\,DLW \qquad (11)$$

where M_0 has the unit of dyne-cm. From the conversion of the seismic moment release (M_0) into the moment magnitude (M_w) in (9) after KANAMORI (1977), the relation of stress drop ($\Delta\tau$) of underthrusting earthquakes in (10), and the definition of the seismic moment release measurement in (11), we can express the earthquake coupling force F_{ec}, in terms of the moment magnitude and the length of the rupture zone

$$F_{ec} = \Delta\tau W = \frac{8}{\pi L W}\,10^{(1.5M_w + 16.05)}. \qquad (12)$$

F_{ec} reflects the degree of seismic coupling at interplate contact. It also indicates the temporal variations in the tectonic coupling F_{tc} within earthquake cycles at subduction zones. It is important to assess the relative size of F_{tc} and F_{ec}.

We estimate L, W, and M_w for each subduction zone from LAY et al. (1982), and TAJIMA and KANAMORI (1985). Comparing the F_{tc} (Table 4) with the F_{ec} (Table 5), we find that about 1% of the F_{tc} contributes to great underthrusting earthquakes in Tonga and the New Hebrides, and 10% of the F_{tc} in the Kuriles and Kamchatka. This result suggests that at most 10% of the F_{tc} may be perturbed by great underthrusting earthquakes, thus the stresses in the plate can also be altered by no more than 10%. The implication is that the major component of F_{tc} appears to remain stable at the interplate contact over geologic time. If great underthrusting earthquakes only perturb the stress field in the plate near the trench by 10%, the neutral surface may not be sufficiently perturbed in depth to locally reverse the earthquake focal mechanism type. Hence, it may be possible to estimate an accurate tectonic coupling force based on the outer rise earthquake data available over a short time window, as we have attempted here. This conclusion seems to agree with the result by WILLEMANN (1991): he used a stress diffusion model to calculate the stress change in the subducting slab at intermediate depth due to shallow underthrusting earthquakes. His result suggests that shallow interplate earthquakes do not reverse the mean stress at the depths where focal mechanisms are observed to change from P–down-dip to T–down-dip orientation. KAGAN and JACKSON (1991a,b) statistically analyzed different time scale properties of several earthquake catalogues and found both short- and long-term clustering for earthquakes of all depth ranges. Such characteristics of earthquake clustering indicate that strong earthquakes are unlikely to release large fractions of the stresses on the fault planes. We see our results as being consistent with what Kagan and Jackson predict for the stress release on fault planes.

Our results predict less than 10% change of stress due to shallow thrust earthquakes. However this conclusion does not exclude the possibility that fluctuations in outer rise and intermediate depth earthquake activities may be temporally related to great underthrusting earthquakes, if the stress in the plate is close to failure in those regions before or after the underthrusting earthquakes. CHRISTENSEN and RUFF (1988) found such correlations at several strongly coupled subduction zones. It is particularly convincing that great earthquakes like 1960 Chile and 1964 Alaska induce an increased number of outer rise extensional events. Recent work by DMOWSKA and LOVISON (1992) shows further evidence of such correlation for the Rat-Island 1965, Alaska 1986, and Valparaiso 1985 great underthrusting earthquakes.

If the stress condition in the plate is close to failure, it will be sensitive to the stress perturbation caused by great underthrusting earthquakes. How sensitive depends on the precise stress change in the plate due to the underthrusting earthquakes. We expect temporal variations in seismic activity to be more significant in regions where the ratio of seismic coupling to tectonic coupling is larger. To examine this hypothesis, we compare seismicity at the four subduction zones studied, where higher tectonic coupling in the plate is found in the New Hebrides and Tonga which have less seismic coupling, and lower tectonic coupling is in Kamchatka and the Kuriles which have higher seismic coupling.

If great underthrusting earthquakes modulate the static stresses in the slab at all, we may expect down-dip tensional earthquakes at intermediate depth and compressional outer rise events before great earthquakes, and down-dip compressional events at intermediate depth and tensional outer rise events after great earthquakes. We examine the intermediate and outer rise earthquakes in the subducting slab above 150 km depth within 10 years before and after the great interplate thrust earthquakes. LAY et al. (1989) made an extensive collection and study of large intraplate earthquakes in coupled subduction zones up to 1986; we will use their data for this comparison, and summarize some observations.

There is lack of great underthrusting earthquakes in the northern segment of the Tonga trench, where most of our outer rise earthquakes are located. The nearest underthrusting earthquake is the 1982 ($M_w = 7.5$) event, which is located at the southern end of the Tonga subduction zone. Seismicity near the 1982 interplate event also shows a complex pattern (CHRISTENSEN and LAY, 1988). There were both compressional and tensional outer rise events prior to the earthquake. The down-dip compressional events were dominant both before and after the 1982 interplate event. Some down-dip tear faulting events are believed to be associated with the subduction of the Louisville Ridge (ISACKS and MOLNAR, 1969; GIARDINI and WOODHOUSE, 1984; CHRISTENSEN and LAY, 1988). Thus, there may not be significant temporal correlation of outer rise and intermediate earthquakes to interplate events, other than possibly for the 1982 event (LAY et al., 1989).

In the New Hebrides, the seismicity pattern shows a mixture of down-dip tensional and compressional earthquakes before and after the 1980 Santa Cruz ($M_w = 7.5$, 7.8) interplate events. One outer rise tensional event occurred afterwards. There was also a mixture of down-dip tensional/compressional foreshocks and aftershocks of the 1973 ($M_s = 7.9$) interplate event, along with several tear faulting events. Down-dip tensional events occurred before and after the 1965 ($M_w = 7.6$) interplate event. Two tensional outer rise events followed that event several years later. Overall, a significant temporal seismic pattern related to the largest underthrusting earthquakes at New Hebrides is not seen.

LAY et al. (1989) found a more significant temporal seismic pattern for the 1963 Kurile Island ($M_w = 8.5$) earthquake. This earthquake was preceded by several large down-dip tensional events, and followed by several tensional outer rise events. There was also a couple of small down-dip compressional events following the main underthrusting event, mixed with one tensional event and one tear event. Intraplate activity near the 1969 Kurile event ($M_w = 8.2$) rupture zone has a general pattern similar to that in the 1963 rupture zone. Several down-dip tensional events occurred prior to the interplate event, and two large outer rise events occurred afterwards. But there were also some tear events and some small down-dip compressional events before and after the mainshock.

Two down-dip compressional events occurred after the 1952 ($M_w = 9.0$) earthquake at Kamchatka (LAY et al., 1989). Following that there was nearly 10 years of quiescence in the vicinity of the rupture zone. Within the next 10 years there were a large tensional outer rise event and many small down-dip compressional events. Several down-dip tensional events occurred below the down-dip compressional events, suggesting the existence of a double Benioff zone there. It seems that a temporal seismicity pattern is significant near Kamchatka. In their global analysis, LAY et al. (1989) find that even when there is evidence for a temporal pattern there is no example of a clear reversal in stress state at a given depth or position in the plate in the outer rise or at intermediate depth that would require a large perturbation of the neutral surface. While this may be a data sampling limitation, it does not preclude the temporal stress perturbations from being small relative to the static stresses.

From the above discussion, the seismicity near large underthrusting earthquakes seems to suggest that the existence of any temporal pattern is correlated with the ratio of F_{ec} to F_{tc} (interplate coupling change). The intensity of outer rise and down-dip activity varies more in regions with a larger coupling change, like the

Kuriles and Kamchatka; whereas for the New Hebrides and Tonga we do not see clear correlations. The complex temporal seismic pattern at the New Hebrides has been attributed to the steep dip of the subducting slab, and the abundance of tear faults (LAY *et al.*, 1989). The complexity of seismicity near the 1982 Tonga earthquake has been considered as the consequence of the subduction of the Louisville Ridge (ISACKS and MOLNAR, 1969; GIARDINI and WOODHOUSE, 1984; CHRISTENSEN and LAY, 1989). Our results suggest that in addition to these considerations, another factor, the background stress, may also play an important role in defining the seismic activities. Of course, the largest earthquakes in each region vary substantially, and it may simply be that temporal patterns only accompany events with particularly large strain changes, such as the great events in the Kurile-Kamchatka arc (LAY *et al.*, 1989).

Figure 12 explores how the tectonic coupling F_{tc} varies with the plate age and convergence rate. We do not see any significant correlation of F_{tc} with either one of the variables, although they are considered significantly correlated with the seismic coupling (RUFF and KANAMORI, 1980). It is too early to draw any meaningful conclusion for the F_{tc} correlations, because our data set is rather limited.

One final consideration is that our models suggest earthquake stress drops are small compared to total stresses in subduction zones. This is, of course, a long standing question, but recent ideas about rupture mechanics (e.g., HEATON, 1990) suggest mechanisms by which rupture can occur under those conditions and satisfy seismic and thermal observations. There may be little problem in accepting such rupture mechanisms given the thermal calculations of MOLNAR and ENGLAND (1990), which indicate high fault stress levels.

Conclusions

We use a model of a 2-D elastic bending plate floating on a "weak" mantle to calculate the stresses in the oceanic lithosphere near the ocean trenches. We perform waveform modeling on outer rise earthquakes to obtain good depths which, along with the focal mechanisms, are used to estimate interplate coupling at subduction zones. We apply a penalty function scheme and F-test likelihood estimation to the search for mean and standard deviation of the tangential load $\bar{f}_x$. We further estimate the tectonic coupling F_{tc} by correcting for slab pull and mantle shear forces. Our results suggest that:

(1) For the four tectonically coupled subduction zones: Kurile, Kamchatka, Tonga, and New Hebrides, the mean tangential loads $\bar{f}_x$ are found to be 166–671 bar for coupling widths of 125–200 km. The associated tectonic coupling force F_{tc} is on the order of 10^{12}–10^{13} N/m, corresponding to 300–600 bar mean shear stress over 200 km coupling width at the interplate contact.

(2) Such tectonic coupling F_{tc} is about one to two orders of magnitude larger than the earthquake coupling F_{ec} associated with great underthrusting earthquakes. The temporal change of F_{tc} is likely to be less than 10% of the total F_{tc} at the interplate contact.

(3) The tectonic coupling F_{tc} is found to be about 6×10^{12} N/m for the Kurile and Kamchatka regions, and about 12×10^{12} N/m for Tonga and the New Hebrides. The ratio of F_{tc} to F_{ec} is greater by a factor of 10 for the Kuriles and Kamchatka than for Tonga and the New Hebrides. The difference of this ratio seems to be compatible with the temporal intraplate seismicity patterns (LAY et al., 1989) showing a relatively more significant correlation with the occurrence of large underthrusting earthquakes for regions with lower F_{tc} than for regions with higher F_{tc}.

(4) The tectonic coupling in the Marianas is low, no greater than 2.4×10^{12} N/m, and is possibly around zero. This result appears to be consistent with the Marianas being a weakly coupled subduction zone.

Appendix 1
Fitting of Synthetic Seismograms to Data

Comparisons of synthetic seismograms and observed seismograms are shown for the five outer rise earthquakes (12/15/80, 02/28/82, 05/11/83 (Figure 2), Tonga; 12/16/85, New Hebrides; 08/23/81, Kamchatka). These earthquakes are all events with refined depth estimates that differ by more than 10 km from the depths given by other studies. The residual of the waveform mismatch is determined by

$$\mathrm{res} = \frac{1}{m} \sum_{i=1}^{m} \left[\frac{\sum_{j=1}^{n} [O_{ij} - (\Delta a_i S_{ij} + c_i)]^2}{\sum_{j=1}^{n} O_{ij}^2} \right]$$

where O is observed seismogram and S is the synthetic seismogram. j is summed over the time series of each seismic record and i is summed over all the stations. Δa_i and c_i are scaling factors for the synthetic seismogram which achieve a best fit to the observed seismogram, and are determined by the smallest residual in the search of depth and duration of the source time function.

Solid lines are observed seismograms and dashed lines are synthetic seismograms. The number shown on the left of each record is the maximum ground displacement in μm. On the right of each record, we plot the station name, symbols to indicate vertical (LPZ) or transverse (LPT) component of the seismogram, epicentral distance (Δ), and station azimuth (Az). Δa and c are the scaling factors defined above. The residual is plotted against the depth in the upper right corner of each figure.

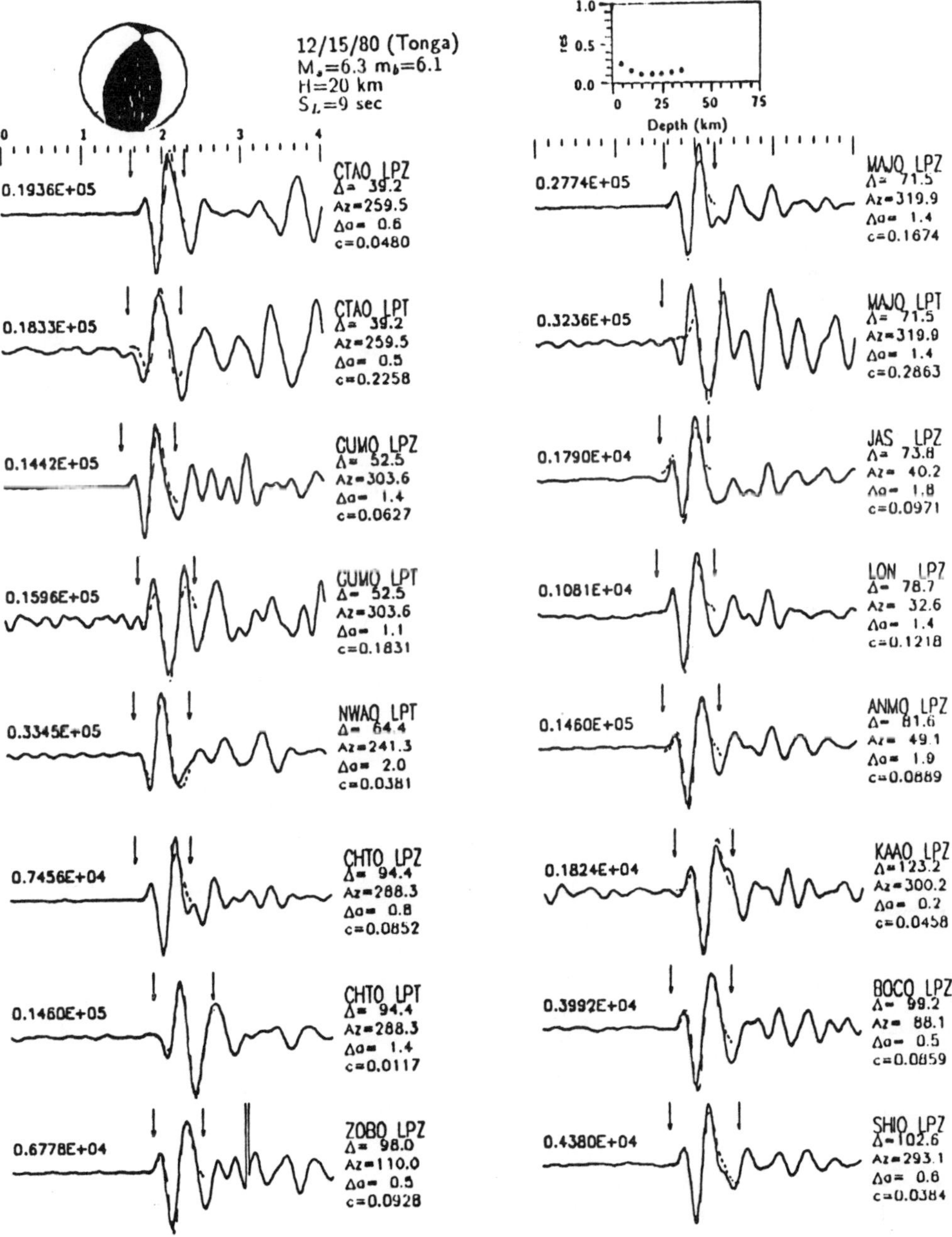
12/15/80 (Tonga)
M_s=6.3 m_b=6.1
H=20 km
S_L=9 sec
Depth (km)
res
1.0
0.5
0.0
0 25 50 75
0 1 2 3 4
0.1936E+05 CTAO LPZ Δ= 39.2 Az=259.5 Δa= 0.6 c=0.0480
0.1833E+05 CTAO LPT Δ= 39.2 Az=259.5 Δa= 0.5 c=0.2258
0.1442E+05 CUMO LPZ Δ= 52.5 Az=303.6 Δa= 1.4 c=0.0627
0.1596E+05 CUMO LPT Δ= 52.5 Az=303.6 Δa= 1.1 c=0.1831
0.3345E+05 NWAO LPT Δ= 64.4 Az=241.3 Δa= 2.0 c=0.0381
0.7456E+04 CHTO LPZ Δ= 94.4 Az=288.3 Δa= 0.8 c=0.0852
0.1460E+05 CHTO LPT Δ= 94.4 Az=288.3 Δa= 1.4 c=0.0117
0.6778E+04 ZOBO LPZ Δ= 98.0 Az=110.0 Δa= 0.5 c=0.0928
0.2774E+05 MAJO LPZ Δ= 71.5 Az=319.9 Δa= 1.4 c=0.1674
0.3236E+05 MAJO LPT Δ= 71.5 Az=319.9 Δa= 1.4 c=0.2863
0.1790E+04 JAS LPZ Δ= 73.8 Az= 40.2 Δa= 1.8 c=0.0971
0.1081E+04 LON LPZ Δ= 78.7 Az= 32.6 Δa= 1.4 c=0.1218
0.1460E+05 ANMO LPZ Δ= 81.6 Az= 49.1 Δa= 1.9 c=0.0889
0.1824E+04 KAAO LPZ Δ=123.2 Az=300.2 Δa= 0.2 c=0.0458
0.3992E+04 BOCO LPZ Δ= 99.2 Az= 88.1 Δa= 0.5 c=0.0859
0.4380E+04 SHIO LPZ Δ=102.6 Az=293.1 Δa= 0.6 c=0.0384

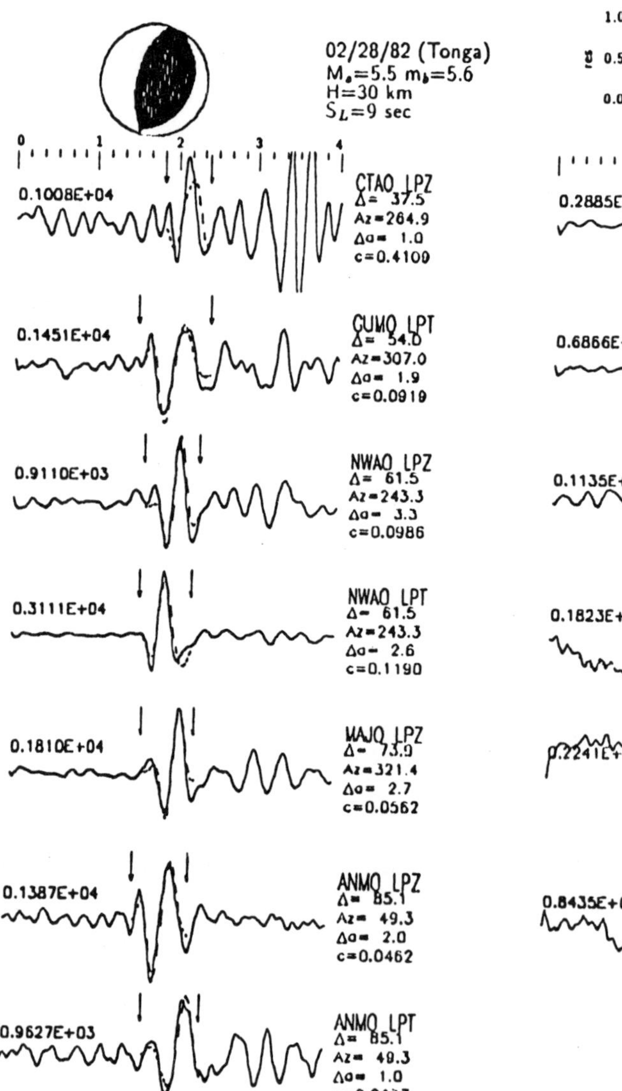
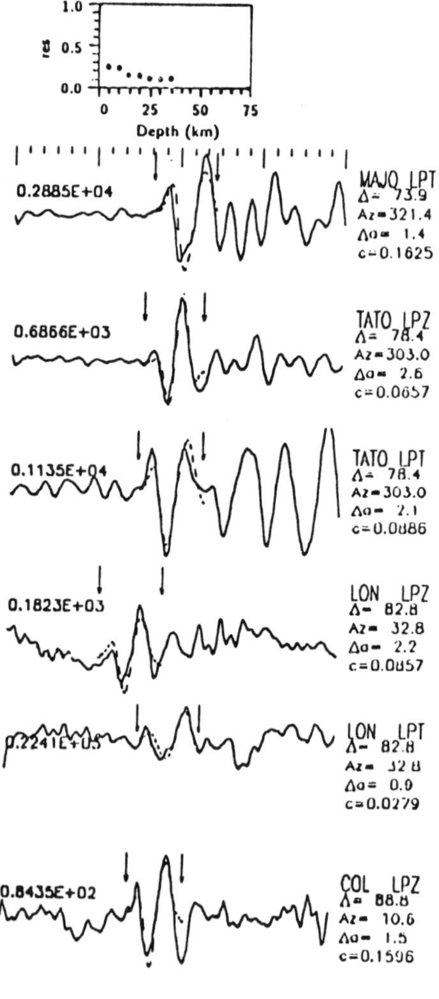
02/28/82 (Tonga)
M_s=5.5 m_b=5.6
H=30 km
S_L=9 sec

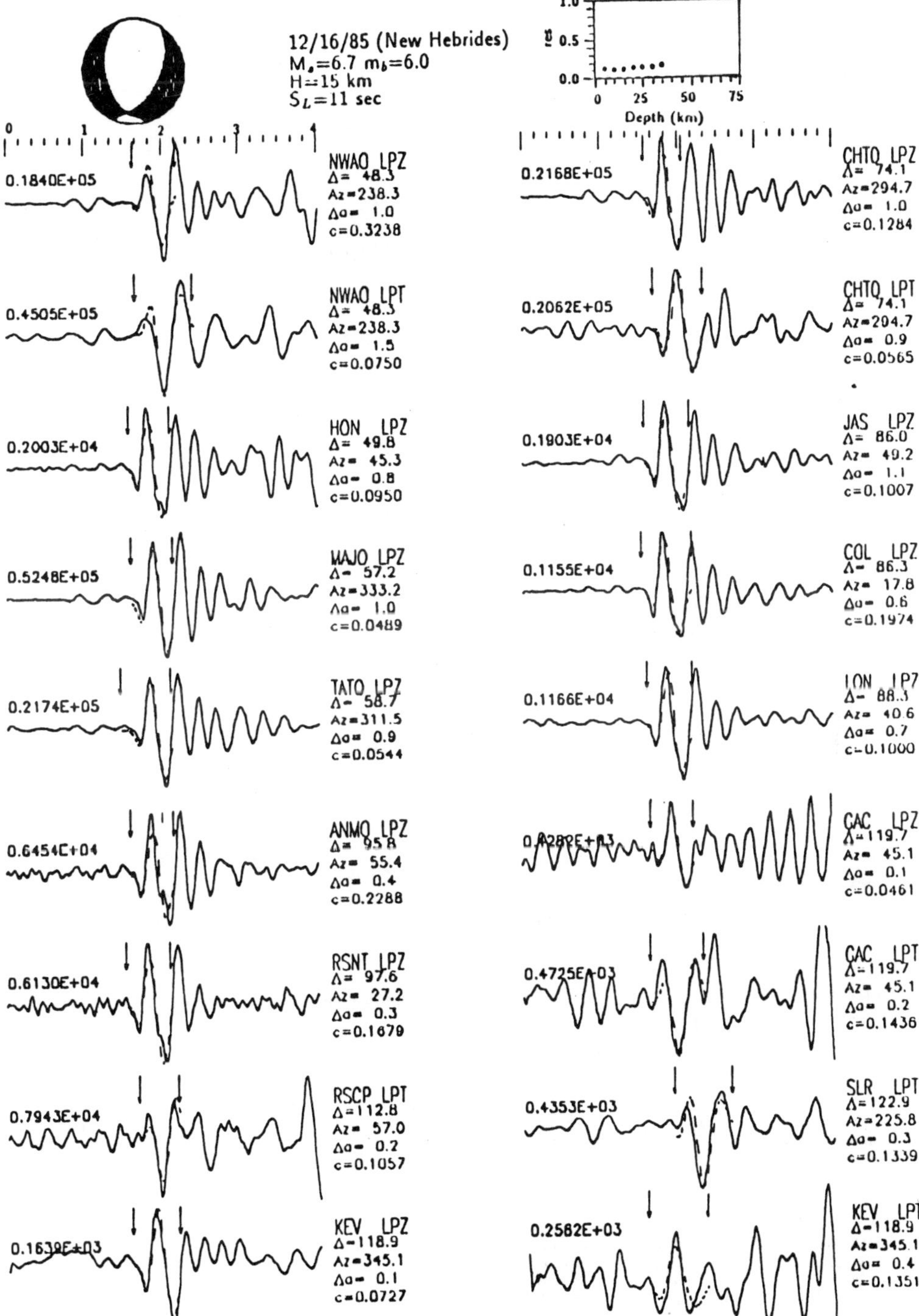

12/16/85 (New Hebrides)
M_s=6.7 m_b=6.0
H=15 km
S_L=11 sec

res
1.0
0.5
0.0
0 25 50 75
Depth (km)

0.1840E+05
NWAO LPZ
Δ= 48.3
Az=238.3
Δa= 1.0
c=0.3238

0.4505E+05
NWAO LPT
Δ= 48.3
Az=238.3
Δa= 1.5
c=0.0750

0.2003E+04
HON LPZ
Δ= 49.8
Az= 45.3
Δa= 0.8
c=0.0950

0.5248E+05
MAJO LPZ
Δ= 57.2
Az=333.2
Δa= 1.0
c=0.0489

0.2174E+05
TATO LPZ
Δ= 58.7
Az=311.5
Δa= 0.9
c=0.0544

0.6454E+04
ANMO LPZ
Δ= 95.8
Az= 55.4
Δa= 0.4
c=0.2288

0.6130E+04
RSNT LPZ
Δ= 97.6
Az= 27.2
Δa= 0.3
c=0.1679

0.7943E+04
RSCP LPT
Δ=112.8
Az= 57.0
Δa= 0.2
c=0.1057

0.1639E+03
KEV LPZ
Δ=118.9
Az=345.1
Δa= 0.1
c=0.0727

0.2168E+05
CHTO LPZ
Δ= 74.1
Az=294.7
Δa= 1.0
c=0.1284

0.2062E+05
CHTO LPT
Δ= 74.1
Az=294.7
Δa= 0.9
c=0.0565

0.1903E+04
JAS LPZ
Δ= 86.0
Az= 49.2
Δa= 1.1
c=0.1007

0.1155E+04
COL LPZ
Δ= 86.3
Az= 17.8
Δa= 0.6
c=0.1974

0.1166E+04
LON LPZ
Δ= 88.3
Az= 40.6
Δa= 0.7
c=0.1000

0.8280E+03
CAC LPZ
Δ=119.7
Az= 45.1
Δa= 0.1
c=0.0461

0.4725E+03
CAC LPT
Δ=119.7
Az= 45.1
Δa= 0.2
c=0.1436

0.4353E+03
SLR LPT
Δ=122.9
Az=225.8
Δa= 0.3
c=0.1339

0.2582E+03
KEV LPT
Δ=118.9
Az=345.1
Δa= 0.4
c=0.1351

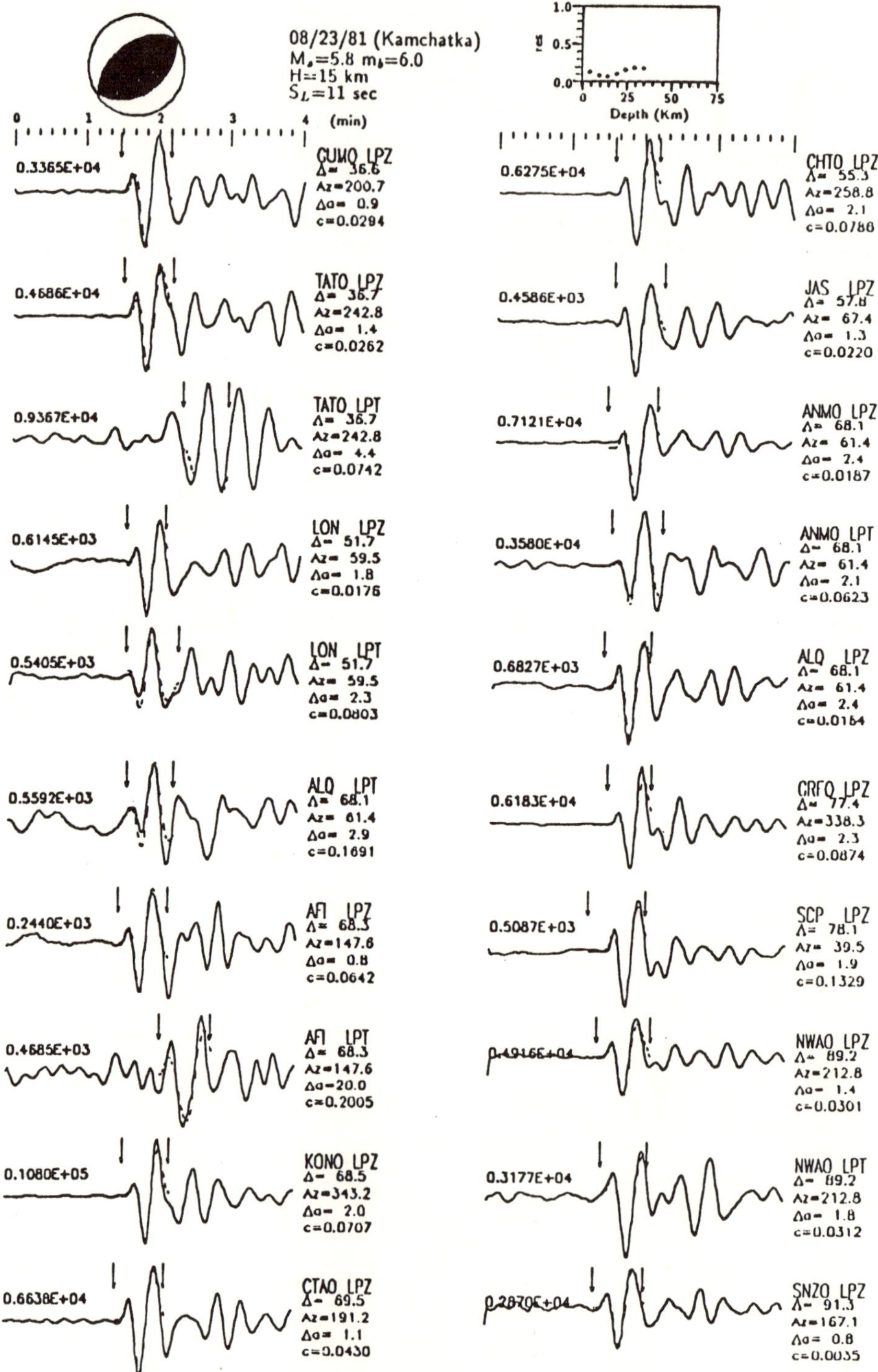
08/23/81 (Kamchatka)
M_s=5.8 m_b=6.0
H=15 km
S_L=11 sec
1.0
res 0.5
0.0
0 25 50 75
Depth (Km)

0   1   2   3   4 (min)

0.3365E+04
CUMO LPZ
Δ= 36.6
Az=200.7
Δa= 0.9
c=0.0294

0.6275E+04
CHTO LPZ
Δ= 55.3
Az=258.8
Δa= 2.1
c=0.0788

0.4686E+04
TATO LPZ
Δ= 36.7
Az=242.8
Δa= 1.4
c=0.0262

0.4586E+03
JAS LPZ
Δ= 57.8
Az= 67.4
Δa= 1.3
c=0.0220

0.9367E+04
TATO LPT
Δ= 36.7
Az=242.8
Δa= 4.4
c=0.0742

0.7121E+04
ANMO LPZ
Δ= 68.1
Az= 61.4
Δa= 2.4
c=0.0187

0.6145E+03
LON LPZ
Δ= 51.7
Az= 59.5
Δa= 1.8
c=0.0176

0.3580E+04
ANMO LPT
Δ= 68.1
Az= 61.4
Δa= 2.1
c=0.0623

0.5405E+03
LON LPT
Δ= 51.7
Az= 59.5
Δa= 2.3
c=0.0803

0.6827E+03
ALQ LPZ
Δ= 68.1
Az= 61.4
Δa= 2.4
c=0.0164

0.5592E+03
ALQ LPT
Δ= 68.1
Az= 61.4
Δa= 2.9
c=0.1691

0.6183E+04
CRFO LPZ
Δ= 77.4
Az=338.3
Δa= 2.3
c=0.0874

0.2440E+03
AFI LPZ
Δ= 68.5
Az=147.6
Δa= 0.8
c=0.0642

0.5087E+03
SCP LPZ
Δ= 78.1
Az= 39.5
Δa= 1.9
c=0.1329

0.4685E+03
AFI LPT
Δ= 68.5
Az=147.6
Δa=20.0
c=0.2005

0.4916E+04
NWAO LPZ
Δ= 89.2
Az=212.8
Δa= 1.4
c=0.0301

0.1080E+05
KONO LPZ
Δ= 68.5
Az=343.2
Δa= 2.0
c=0.0707

0.3177E+04
NWAO LPT
Δ= 89.2
Az=212.8
Δa= 1.8
c=0.0312

0.6638E+04
CTAO LPZ
Δ= 69.5
Az=191.2
Δa= 1.1
c=0.0430

0.2870E+04
SNZO LPZ
Δ= 91.3
Az=167.1
Δa= 0.8
c=0.0035

Appendix 2
Numerical Calculations of the Stresses in the Outer Rise

To approximate conditions in the oceanic lithosphere during earthquake cycles, we employ a two-demensional, isotropic, layer-over-halfspace model. More complicated models seem unnecessary because earthquake data do not provide complete constraints on the stresses. The time transformed quasi-static equations of motion in a (as yet) nongravitating layered media is

$$\nabla \cdot \mathbf{t}(\mathbf{r}, \lambda(\omega), \mu(\omega), \omega) = \mathbf{0} \tag{1}$$

where the stress tensor

$$\mathbf{t}(\mathbf{r}, \omega) = \lambda(\mathbf{r}, \omega) \, \nabla \cdot \mathbf{u}(\mathbf{r}, \omega)\mathbf{I} + \mu(\mathbf{r}, \omega)[\nabla\mathbf{u}(\mathbf{r}, \omega) + \nabla^T\mathbf{u}(\mathbf{r}, \omega)]; \tag{2}$$

here ω is frequency. $\mathbf{r}$ is a position vector of x, y and z in a Cartesian frame. λ is Lamé constant. μ is rigidity. $\mathbf{I}$ is an identity matrix. $\mathbf{u}$ is a displacement vector. Equation (1) is to be solved subject to certain boundary conditions (possibly time dependent) in the halfspace and at the top of the layer. In (1), the Lamé parameters are functions of frequency ω, appropriate for any linear viscoelastic rheology. x axis is chosen to be horizontally orthogonal to the strike of the trench, positive towards the ocean. y axis is horizontally parallel to the strike of the trench. And z axis is positive downward. We consider a 2-D problem that is independent of the y coordinate. Solutions to (1) are particularly easy to formulate using propagator matrices (WARD, 1984). The technique reduces two second-order equations in displacement to four first-order equations in displacement and stress. The latter are much easier to propagate vertically through a layered media. A wavenumber transform on the x dependence of (1)

$$f(k, z, \omega) = \int_{-\infty}^{+\infty} f(x, z, \omega) \, e^{-ikx} \, dx \tag{3}$$

yields the first-order equations

$$\frac{\partial}{\partial z} \mathbf{v}(k, z, \omega) = \mathbf{M}(k, z, \omega)\mathbf{v}(k, z, \omega) \tag{4}$$

where $\mathbf{v}$ is the stress-displacement vector

$$\mathbf{v}(k, z, \omega) = \begin{bmatrix} iu_x(k, z, \omega) \\ u_z(k, z, \omega) \\ it_{xz}(k, z, \omega) \\ t_{zz}(k, z, \omega) \end{bmatrix} \tag{5}$$

and the elements of $\mathbf{M}$ are

$$\mathbf{M}(k, z, \omega) = \begin{bmatrix} 0 & k & \mu^{-1} & 0 \\ \dfrac{-\lambda k}{(\lambda + 2\mu)} & 0 & 0 & \dfrac{1}{(\lambda + 2\mu)} \\ \dfrac{4\mu k^2(\lambda + \mu)}{(\lambda + 2\mu)} & 0 & 0 & \dfrac{\lambda k}{(\lambda + 2\mu)} \\ 0 & 0 & -k & 0 \end{bmatrix}. \tag{6}$$

The general solution to (4) is

$$\mathbf{v}(k, z, \omega) = C_1 \mathbf{e}^- e^{-kz} + C_2(\mathbf{u}^- + z\mathbf{e}^-) e^{-kz} + C_3 \mathbf{e}^+ e^{kz} + C_4(\mathbf{u}^+ + z\mathbf{e}^+) e^{kz} \tag{7}$$

where C_1, C_2, C_3, and C_4 are to be determined by the boundary conditions. $\mathbf{e}^+$, $\mathbf{u}^+$, $\mathbf{e}^-$, and $\mathbf{u}^-$ are the eigenvectors of $\mathbf{M}(k, z, \omega)$ corresponding to the eigenvalues of k and $-k$

$$\mathbf{e}^- = (1, 1, -2\mu k, -2\mu k)^T$$

$$\mathbf{u}^- = \left(-\frac{\lambda + 2\mu}{k(\lambda + \mu)}, \frac{\mu}{k(\lambda + \mu)}, 2\mu, 0\right)^T \tag{8}$$

$$\mathbf{e}^+ = (-1, 1, -2\mu k, 2\mu k)^T$$

$$\mathbf{u}^+ = \left(-\frac{\lambda + 2\mu}{k(\lambda + \mu)}, \frac{-\mu}{k(\lambda + \mu)}, -2\mu, 0\right)^T.$$

Solutions for the unknown displacements and stresses through propagation can be expressed as

$$\mathbf{v}(k, z, \omega) = \mathbf{P}(k, z, z_0, \omega)\mathbf{v}(k, z_0, \omega). \tag{9}$$

The propagator matrix $\mathbf{P}(k, z, z_0, \omega)$ carries solutions $\mathbf{v}$ at the top of the halfspace ($z = z_0$) upward to depth z in the layer. We will restrict attention to homogeneous layered medium, so within a layer, $\lambda(z, \omega) \to \lambda(\omega)$, $\mu(z, \omega) \to \mu(\omega)$. The propagator can be written in terms of the eigenvalues $(k, -k)$ and eigenvectors $(\mathbf{e}^+, \mathbf{e}^-, \mathbf{u}^+, \mathbf{u}^-)$ of $\mathbf{M}(k, z, \omega) \to \mathbf{M}(k, \omega)$

$$\mathbf{P}(k, z, z_0, \omega) = [\mathbf{e}^-, \mathbf{u}^-, \mathbf{e}^+, \mathbf{u}^+]$$

$$\times \begin{bmatrix} e^{-k\Delta} & \Delta e^{-k\Delta} & 0 & 0 \\ 0 & e^{-k\Delta} & 0 & 0 \\ 0 & 0 & e^{k\Delta} & \Delta e^{k\Delta} \\ 0 & 0 & 0 & e^{k\Delta} \end{bmatrix} [\mathbf{e}^-, \mathbf{u}^-, \mathbf{e}^+, \mathbf{u}^+]^{-1} \tag{10}$$

where $\Delta = z - z_0$.

Boundary Conditions

Four boundary conditions are needed to specify a unique solution to (9). Two conditions will be applied at the top of the halfspace and two will be applied at the top of the oceanic plate. In the halfspace, $\mathbf{v}(k, z, \omega)$ in (7) must decay with depth, thus, at the bottom of the lithosphere,

$$\mathbf{v}(k, z_0, \omega) = A(k, \omega)\mathbf{e}^-(k, z_0, \omega) + B(k, \omega)\mathbf{u}^-(k, z_0, \omega). \tag{11}$$

At the top of the oceanic lithosphere we assume

$$\mathbf{v}(k, z = 0, \omega) = [C(k, \omega), u_z(k, z = 0, \omega), it_{xz}(k, z = 0, \omega), D(k, \omega)]^T \tag{12}$$

where u_z and t_{xz} are specified functions. The horizontal shear stress, t_{xz}, will play the role of variable interplate coupling. The vertical displacement of the plate u_z is composed of the trench-outer rise topography and down-dip plate shape estimated from the seismicity. We solve equations (9), (11), and (12) for four unknowns $A(k, \omega)$, $B(k, \omega)$, $C(k, \omega)$ and $D(k, \omega)$ at all k and ω. The stresses and displacements at any depth can be obtained by propagating the bottom solution $\mathbf{v}(k, z_0, \omega)$ in (11). $D(k, \omega)$ is the vertical loads applied on the surface of the plate given plate thickness, rigidity and vertical displacement.

Gravity

Gravity terms ignored in the equation of motion (1) are of the order $\rho g/\mu k$ relative to the elastic terms. The influence of gravity is smaller in coseismic stress modeling of typical earthquakes since μk is commonly much greater than ρg. However, in modeling interseismic deformation and the whole earthquake cycle, gravity terms cannot be dropped since the wavelengths involved are considerably longer and rigidities are much smaller due to relaxation effects. Gravity effect is taken into account by adding an advected pressure to the stresses resulting from elastic deformation (WARD, 1984). An extra increment of pressure $PI = -\rho g u_z I$ appears in the stress tensor

$$\mathbf{t} = \lambda\theta\mathbf{I} + 2\mu\epsilon + P\mathbf{I} \tag{13}$$

where $\theta = \nabla \cdot \mathbf{u}$ and ϵ is a strain vector. If a new stress-displacement vector is formed as $\hat{\mathbf{v}}(k, z, \omega) = [iu_x, u_z, it_{xz}, t_{zz} - \rho g u_z]^T$ then $\hat{\mathbf{v}}(k, z, \omega) = \mathbf{G}\mathbf{v}(k, z, \omega)$ with

$$\mathbf{G} = \begin{bmatrix} 1 & 0 & 0 & 0 \\ 0 & 1 & 0 & 0 \\ 0 & 0 & 1 & 0 \\ 0 & -\rho g & 0 & 1 \end{bmatrix}. \tag{14}$$

Within a homogeneous layer, $\hat{\mathbf{v}}(k, z, \omega) = \hat{\mathbf{P}}(k, z, z_0, \omega)\hat{\mathbf{v}}(k, z_0, \omega)$, so the propagator which includes advected pressure is $\hat{\mathbf{P}}(k, z, z_0, \omega) = \mathbf{G}\mathbf{P}(k, z, z_0, \omega)\mathbf{G}^{-1}$. By substituting $\hat{\mathbf{P}}(k, z, z_0, \omega)$ for $\mathbf{P}(k, z, z_0, \omega)$ and replacing vector $\mathbf{v}(k, z_0, \omega)$ by $\hat{\mathbf{v}}(k, z_0, \omega)$ gravity is accounted for.

Pseudo Time Dependence

The theoretical formulation so far is time dependent (through varying boundary conditions and Lamé constants) and completely applicable to any linear viscoelastic rheology. Solutions in the space and time domain are obtained from (9) by the double inverse transform

$$\mathbf{v}(x, z, t) = (2\pi)^{-2} \int_{-\infty}^{\infty} e^{-i\omega t} \, d\omega \int_{-\infty}^{\infty} \mathbf{v}(k, z, \omega) \, e^{ikx} \, dk. \tag{15}$$

Explicit time dependent behavior of the stresses depends on the viscoelastic composition of the lithosphere and mantle. Because of the highly underdetermined nature of the problem, we consider only a simple stress approximation. Viscoelastic relaxation effect can be approximated by considering the response of a relaxed earth. For the forces applied on the subducting slab for duration longer than the mantle relaxation time, relaxed (geologic) rigidity is used. We only allow shear stresses to relax. Lamé constant λ remains the same as the seismic Lamé constant since it does not vary largely in time. The modeling normal load (normal stress on top of the entire plate) represents overburden pressure, slab pull component normal to plate surface, and possibly induced mantle flow pressure. The modeling tangential load (shear stress over a width of maximal 200 km at the interplate contact) simulates the total accumulated shear stresses on the surface of the entire subducting slab. The tangential load bears the information of the mechanical coupling at the interplate contact. Elastic responses of a plate to the normal and tangential loads are added together to form the stresses in the bending plate (Figure 3(a1) and (a2)).

Acknowledgments

We thank Steven Ward for helpful discussions on this study, especially for his contribution of suggesting that stresses in the plate could be partitioned into a combination of those resulting from normal and tangential loads. We thank Heidi Houston, Thorne Lay, Zhengkang Shen, Jiajun Zhang, and Xiao-Bi Xie for their helpful discussions, suggestions, comments, and criticisms on the manuscript. We also thank two anonymous reviewers for their suggestions and comments which improved the paper. This research was partially supported by the National Science Foundation grants EAR–8904707 (K. C. McNally); EAR–8507555, EAR–8618222 (S. N. Ward), and NASA grant NAG–5-750 (S. N. Ward) with facilities supported from the W. M. Keck Foundation. Contribution 162, C. F. Richter Seismological Laboratory, Institute of Tectonics, University of California, Santa Cruz.

REFERENCES

BEVINGTON, P. R., *Data Reduction and Error Analysis for the Physical Sciences* (McGraw-Hill Book Company, 1969).

BURBACH, G. V., and FROHLICH, C. (1986), *Intermediate and Deep Seismicity and Lateral Structure of Subducted Lithosphere in the Circum-Pacific Region*, Rev. Geophys. *24*, 833–874.

CALDWELL, J. G., HAXBY, W. F., KARIG, D. E., and TURCOTTE, D. L. (1976), *On the Applicability of a Universal Elastic Trench Profile*, Earth Planet. Sci. Lett. *31*, 239–246.

CATHLES, L. M. III, *The Viscosity of the Earth's Mantle* (Princeton University Press, 1975).

CHAPPLE, W. M., and FORSYTH, D. W. (1979), *Earthquakes and Bending of Plates at Trenches*, J. Geophys. Res. *84*, 6729–6749.

CHEN, T., and FORSYTH, D. W. (1978), *A Detailed Study of Two Earthquakes Seaward of the Tonga Trench: Implications for Mechanical Behavior of the Oceanic Lithosphere*, J. Geophys. Res. *83*, 4995–5003.

CHEN, W.-P., and MOLNAR, P. (1983), *Focal Depths of Intracontinental and Intraplate Earthquakes and their Implications for the Thermal and Mechanical Properties of the Lithosphere*, J. Geophys. Res. *88*, 4183–4214.

CHOY, G. L., and DEWEY, J. W. (1988), *Rupture Process of an Extended Earthquake Sequence: Teleseismic Analysis of Chilean Earthquake of March 3, 1985*, J. Geophys. Res. *93*, 1103–1118.

CHRISTENSEN, D. H., and LAY, T. (1988), *Large Earthquakes in the Tonga Region Associated with Subduction of the Louisville Ridge*, J. Geophys. Res. *93*, 13,367–13,389.

CHRISTENSEN, D. H., and RUFF, L. J. (1988), *Seismic Coupling and Outer-rise Earthquakes*, J. Geophys. Res. *93*, 13,421–13,444.

CHRISTENSEN, D. H., and RUFF, L. J. (1983), *Outer Rise Earthquakes and Seismic Coupling*, Geophys. Res. Lett. *10*, 697–700.

DAS, S., and SCHOLZ, C. H. (1983), *Why Large Earthquakes do not Nucleate at Shallow Depths*, Nature *305*, 621–623.

DAVIES, G. F. (1980), *Mechanics of Subducted Lithosphere*, J. Geophys. Res. *85*, 6304–6318.

DMOWSKA, R., and LOVISON, L. C. (1992), *Influence of Asperities along Subduction Interfaces on the Stressing and Seismicity of Adjacent Areas*, Tectonophysics *211*, 23–43.

DMOWSKA, R., RICE, J. R., LOVISON, L. C., and JOSELL, D. (1988), *Stress Transfer and Seismic Phenomena in Coupled Subduction Zones during the Earthquake Cycle*, J. Geophys. Res. *93*, 7869–7884.

DZIEWONSKI, A. M., and ANDERSON, D. L. (1981), *Preliminary Reference Earth Model (PREM)*, Phys. Earth Planet. Inter. *25*, 297–356.

DZIEWONSKI, A. M., EKSTRÖM, G., FRANZEN, J. E., and WOODHOUSE, J. H. (1988), *Global Seismicity of 1981: Centroid-moment Tensor Solutions for 542 Earthquakes*, Phys. Earth Planet. Inter. *50*, 155–182.

FORSYTH, D. W. (1982), *Determinations of Focal Depths of Earthquakes Associated with the Bending of Oceanic Plates at Trenches*, Phys. Earth Planet. Inter. *28*, 141–160.

GIARDINI, D., and WOODHOUSE, J. H. (1984), *Deep Seismicity and Modes of Deformation in Tonga Subduction Zone*, Nature *307*, 505–509.

HASEBE, K., FUJII, N., and UYEDA, S. (1970), *Thermal Processes under Island Arcs*, Tectonophysics *10*, 335–355.

HEATON, T. H. (1990), *Evidence for and Implications of Self-healing Pulses of Slip in Earthquake Rupture*, Phys. Earth Planet. Inter. *64*, 1–20.

ISACKS, B. L., and BARAZANGI, M. (1977), *Geometry of Benioff Zones: Lateral Segmentation and Downwards Bending of Subducted Lithosphere*, Island Arcs, Deep Sea Trenches and Back-arc Basins, Maurice Ewing Series *1*, 99–114.

ISACKS, B. L., and MOLNAR, P. (1969), *Mantle Earthquake Mechanisms and the Sinking of the Lithosphere*, Nature *223*, 1121–1124.

JAEGER, J. C., and COOK, N. G. W., *Fundamentals of Rock Mechanics* (London: Chapman and Hall, 1979).

JARRARD, R. D. (1986), *Relations among Subduction Parameters*, Rev. Geophys. *24*, 217–284.

KAGAN, Y. Y., and JACKSON, D. D. (1991a), *Long-term Earthquake Clustering*, Geophys. J. Int. *104*, 117–133.

KAGAN, Y. Y., and JACKSON, D. D. (1991b), *Seismic Gap Hypothesis: Ten Years After*, J. Geophys. Res. *96*, 21,419–21,431.

KANAMORI, H. (1977), *The Energy Release in Great Earthquakes*, J. Geophys. Res. *82*, 2981–2987.

KELLEHER, J., SAVINO, J., ROWLETT, H., and McCANN, W. (1974), *Why and Where Great Thrust Earthquakes Occur along Island Arcs*, J. Geophys. Res. *79*, 4889–4899.

LAY, T., ASTIZ, L., KANAMORI, H., and CHRISTENSEN, D. H. (1989), *Temporal Variation of Large Intraplate Earthquakes in Coupled Subduction Zones*, Phys. Earth Planet. Inter. *54*, 258–312.

LAY, T., KANAMORI, H., and RUFF, L. J. (1982), *The Asperity Model and the Nature of Large Subduction Zone Earthquakes*, Earthq. Pred. Res. *1*, 3–71.

MOLNAR, P., and ENGLAND, P. (1990), *Temperatures, Heat Flux, and Frictional Stress near Major Thrust Faults*, J. Geophys. Res. *95*, 4833–4856.

RUFF, L. J. (1989), *Do Trench Sediments Affect Great Earthquake Occurrence in Subduction Zones?* Pure and Appl. Geophys. *129*, 263–282.

RUFF, L. J., and KANAMORI, H. (1983), *Seismic Coupling and Uncoupling at Subduction Zones*, Tectonophysics *99*, 99–117.

RUFF, L. J., and KANAMORI, H. (1980), *Seismicity and the Subduction Process*, Phys. Earth Planet. Inter. *23*, 240–252.

SCHOLL, D. W., VAN HUENE, R., and DIEFFENBACK, H. L. (1990), *Rates of Sediment Subduction Erosion—Implications for Growth of Terrestrial Crust* (abs.), EOS, Trans. AGU *71*, 1576.

SCHWARTZ, S. Y., DEWEY, J. W., and LAY, T. (1989), *Influence of Fault Plane Heterogeneity on the Seismic Behavior in the Southern Kurile Islands Arc*, J. Geophys. Res. *94*, 5637–5649.

SPENCE, W. (1987), *Slab Pull and the Seismotectonics of Subducting Lithosphere*, Rev. Geophys. *25*, 55–69.

STAUDER, W. (1973), *Mechanisms and Spatial Distribution of Chilean Earthquakes with Relation to Subduction of Oceanic Plate*, J. Geophys. Res. *78*, 5033–5061.

STAUDER, W. (1968a), *Mechanism of the Rat Island Earthquake Sequence of February 4, 1965, with Relation to Island Arcs and Sea-floor Spreading*, J. Geophys. Res. *73*, 3847–3858.

STAUDER, W. (1968b), *Tensional Character of Earthquake Foci Beneath the Aleutian Trench with Relation to Sea Floor Spreading*, J. Geophys. Res. *73*, 7693–7701.

TAJIMA, F., and KANAMORI, H. (1985), *Global Survey of Aftershock Area Expansion Patterns*, Phys. Earth Planet. Inter. *40*, 77–134.

TOKSÖZ, M. N., MINEAR, J. W., and JULIAN, B. P. (1971), *Temperature Field and Geophysical Effects of a Downgoing Slab*, J. Geophys. Res. *76*, 1113–1138.

TOKSÖZ, M. N., SLEEP, N. H., and SMITH, A. T. (1973), *Evolution of the Downgoing Lithosphere and the Mechanisms of Deep Focus Earthquakes*, Geophys. J. R. Astr. Soc. *35*, 285–310.

TURCOTTE, D. L., McADOO, D. C., and CALDWELL, J. G. (1978), *An Elastic–Perfectly Plastic Analysis of the Bending in the Lithosphere at a Trench*, Tectonophysics *47*, 193–206.

TURCOTTE, D. L., and SCHUBERT, G., *Geodynamics: Applications of Continuum Physics to Geological Problems* (John Wiley and Sons, 1982).

UYEDA, S., and KANAMORI, H. (1979), *Back-arc Opening and the Mode of Subduction*, J. Geophys. Res. *84*, 1049–1061.

VAN DEN BUEKEL, J., and WORTEL, R. (1988), *Thermo-mechanical Modeling of Arc-trench Regions*, Tectonophysics *154*, 177–193.

WARD, S. N. (1984), *A Note on Lithospheric Bending Calculations*, Geophys. J. R. Astr. Soc. *78*, 241–253.

WARD, S. N. (1983), *Body Wave Inversion: Moment Tensors and Depths of Oceanic Intraplate Bending Earthquakes*, J. Geophys. Res. *88*, 9315–9330.

WATTS, A. B., BODINE, J. H., and STECKLER, M. S. (1980), *Observations of Flexure and the State of Stress in the Oceanic Lithosphere*, J. Geophys. Res. *85*, 6369–6376.

WIENS, D. A., and STEIN, S. (1983), *Age Dependence of Oceanic Intraplate Seismicity and Implications for Lithospheric Evolution*, J. Geophys. Res. *88*, 6455–6468.

WILLEMANN, R. J. (1991), *Stress Propagation and Strain Rate in Subducted Lithosphere*, J. Geophys. Res. *96*, 10,219–10,232.

                    Interplate Coupling

YOKOKURA, T. (1981), *Viscosity of the Earth's Mantle: Inference from Dynamic Support by Flow Stress*, Tectonophysics *77*, 35–62.

ZHANG, J., and LAY, T. (1992), *The April 5, 1990 Mariana Islands Earthquake and Subduction Zone Stresses*, Phys. Earth Planet. Inter. *72*, 99–121.

ZHANG, J., HAGER, B. H., and RAEFSKY, A. (1985), *Critical Assessment of Viscous Models of Trench Topography and Corner Flow*, Geophys. J. R. Astr. Soc. *83*, 451–475.

ZHOU, H.-W. (1990), *Observations on Earthquake Stress Axes and Seismic Morphology of Deep Slabs*, Geophys. J. Int. *103*, 377–401.

(Received March 30, 1992, revised December 5, 1992, accepted December 20, 1992)

PAGEOPH, Vol. 140, No. 2 (1993)

0033–4553/93/020257–29$1.50 + 0.20/0
© 1993 Birkhäuser Verlag, Basel

Seismicity in Subduction Zones from Local and Regional Network Observations

CARL KISSLINGER[1]

Abstract—The data provided by local and regional seismograph networks are essential for the solution of many problems of subduction-zone seismology. The capabilities of such networks are limited by the instrumentation currently in common use and the unfavorable source-station geometry often imposed by the regional geography. Nevertheless, important contributions have come from the data gathered in many of the earth's subduction zones. The accuracy of hypocenter locations based on regional data is affected by the complex velocity structures characteristic of subduction zones, but the problems are now well-understood. Examples of numerous studies of the spatial configurations of the seismicity in subduction zones and consequent interpretations of seismogenesis and subduction processes are reviewed. Studies of the distributions of earthquakes in time and with magnitude, for events down to the microearthquake level, have the potential for clarifying the earthquake-generating processes and, possibly, a basis for earthquake prediction. Other uses of local and regional network data have been for investigations of coda-Q and the identification of asperities in subduction zones.

Key words: Network seismology, subduction zone, Wadati-Benioff zone.

Introduction

Active or fossil subduction zones are the sites of most of the earth's great earthquakes, all of the intermediate-depth and deep earthquakes, and a very large fraction of the total seismic energy release. In a general sense, these are also among the best understood earthquakes in terms of the source of the tectonic forces that cause them.

Because of the abundance of seismicity and good general understanding of the operative tectonics, observations of subduction-zone earthquakes offer an excellent base for solving a number of fundamental problems of seismogenesis, internal earth structure, and plate-tectonics physics. Important goals of subduction-zone seismology, and the need for abundant observations at closely spaced sites are suggested by the following statements:

"Various effects of the anomalous structure beneath the Japanese islands on the transmission of seismic waves have been observed by numbers of Japanese

---

[1] Cooperative Institute for Research in Environmental Science and Department of Geological Sciences, Campus Box 216, University of Colorado, Boulder, CO 80309, U.S.A.

seismologists over the past fifty years, thanks to a dense distribution of seismic stations over the islands. . . . it is obvious that the distribution of earthquakes is closely related to the anomalous structure." (UTSU, 1971), and

"Knowledge of the geometry of the inclined seismic zones . . . in the upper mantle beneath island arcs is essential for determination of the location as well as the rheological properties of the descending plates, and the relationship of the descending plates to surface tectonic features." (BARAZANGI and ISACKS, 1979).

One goal of the contemporary research on seismogenesis is understanding the geological factors and ambient physical conditions that control: (1) the time interval between strong earthquakes on a given segment of a subduction zone, (2) the localization of the nucleation point of strong earthquakes, (3) the length and width of the segment of the main thrust zone that ruptures in a particular event, (4) the distribution on the rupture surface of moment release during strong earthquakes, (5) the partitioning of long subduction-type plate boundaries into distinct seismogenic units, (6) the association of moderate and small earthquakes with specific geological and/or topographic features, and (7) total number, distribution and rate of decay of aftershocks.

Except for paleoseismic and geodetic studies in available land areas, field investigations of the kind that are providing important data for seismic zones in continental settings, and which could contribute to the solutions of some of these problems, are feasible only in those subduction zones in which the epicentral regions of the strongest earthquakes are not covered by ocean water. Observations of the abundant small earthquakes are an essential source of data for the investigation of processes in the seismogenic zone. These data may eventually provide a basis for the deterministic prediction of future great earthquakes. Developments in the global seismic observing system have greatly improved the capability to investigate many of these problems, using teleseismically recorded signals. Nevertheless, only densely spaced local and regional networks can provide the high-resolution data required for satisfactory resolution of many of these questions.

The obvious unique contributions of local and regional networks come from the ability to provide an enhanced data base by the detection and location of events to the microearthquake level. The possibility, under favorable circumstances, of locating many events relative to each other to within a few kilometers serves the effort to associate seismicity with particular geological features. On the other hand, as will be discussed, the locations of events at intermediate depths by local data are systematically biased by the effects of the subducting slab and are inferior to teleseismically determined locations (McLAREN and FROHLICH, 1985).

The large number of events in a typical regional catalog is useful for identifying and outlining active structures. The statistical significance of changes with time of the rate of occurrence of earthquakes is increased by the large number of data. Waveforms recorded at small distances are useful for studying the distribution of attenuation, especially by means of coda-Q.

Limitations of the effectiveness of regional networks in subduction zones come partly from the instrumentation in common use and partly from the source-station geometry, especially in island arcs. Many networks were installed for the primary purpose of detecting and locating the regional seismicity to small magnitudes. Stations are mostly equipped with vertical component, short-period, narrow-band seismographs. The solution to the limitations imposed by equipment is readily available and requires only a major investment of funds. A few widely spaced broad-band, high dynamic-range, three-component instruments will provide important data essential for some problems, like the distribution of moment release in big earthquakes. They will not provide the high spatial resolution of earthquake locations needed to solve most of the problems listed above.

The problems resulting from the geometry of deployment arise in those situations in which the abundant seismicity occurs seaward of the land areas available for siting permanent stations. With the hypocenters outside the network and limited azimuthal coverage, well-known problems of location accuracy, especially depth determination, and focal mechanism determination are inevitable. An obvious solution is the use of ocean-bottom seismographs in temporary deployments (LAWTON *et al.*, 1982; FROHLICH *et al.*, 1982) and as permanent installations, as has been done on a limited basis in Japan (SEISMOLOGY AND VOLCANOLOGY RESEARCH DIVISION, MRI, 1980). Again, the wide use of this solution is inhibited by the large capital costs.

In spite of these limitations, important contributions to the understanding of subduction zones and the associated earthquakes have come from the analysis of local and regional network data. Examples of recent results are summarized in this paper. The relevant literature is extensive, so emphasis is given to key papers which are themselves summaries of bodies of work and which contain comprehensive bibliographies. The potential for future advances to arise from improved and more widely deployed networks should be apparent from the results already achieved.

An excellent and comprehensive review of the contributions of regional networks to subduction-zone seismology was published by UTSU (1971). His paper, which includes results from 1918 to the time of publication, offers an extensive bibliography. The fact that the region under and around Japan dominates the review reflects the fact that monitoring of other subduction-type seismic zones by densely spaced regional stations was a recent development at that time. The present review attempts to illustrate by examples what we have learned, especially in regard to the distribution of seismicity and the configuration of subducted plates, in the 20 years since Utsu's paper.

Network Coverage of Subduction Zones

Many subduction zones are now monitored by regional networks, and in a few cases, dense local networks. Networks are occasionally closed, new ones are installed, and sometimes the networks are deployed for a limited time only. The coverage is not guided by a global plan. The quality of the networks and the density of station spacing is far from uniform, but the widespread interest in the resulting data is clear. A general, but incomplete, view of the coverage, based on the reporting of data to the National Earthquake Information Center, U.S. Geological Survey, may be seen in PRESGRAVE *et al.* (1985). Important contributions from arrays of strong-motion seismographs (e.g., CASTRO *et al.*, 1990) and from temporary deployments of mobile stations, such as the PANDA array (CHIU *et al.*, 1991; PUJOL *et al.*, 1991) are not included in this review, except for examples of results from Ocean Bottom Seismograph (OBS) observations.

Subduction locations that are monitored, or have been monitored in recent times, are listed in Table 1. The approximate length of the network parallel to the

Table 1

Some subduction zones monitored by regional or local networks

Site	Operator
Central Aleutian Islands (Adak)	Univ. of Colorado (150 km)
Eastern Aleutian Islands (Shumagin)	Columbia Univ. (260 km)
Southern Mainland Alaska	Univ. of Alaska, U.S. Geological Survey (1400 km)
Cascadia Subduction Zone	Univ. of Washington, U.S.G.S. Western Canada Telemetered Network (1200 km)
Mexico	Univ. Nac. Auton. Mexico (1200 km)
Guatemala	Inst. Naçional de Sismol., Vulcanol., Meterol., e Hidrol. (360 km)
El Salvador	Centro de Investigaçiones Geotecnias (200 km)
Nicaragua	Inst. Nacional de Estudias Territoriales (300 km)
Costa Rica	Inst. Costarricense de Electric./U. Texas, Austin; Univ. Nacional/U. Calif., Santa Cruz (350 km)
Caribbean Sea	U.S.G.S., Columbia University, U. of the West Indies (1300 km)
Peru	Inst. Geofisica Peru (1750 km)
Chile	Univ. de Chile (2400 km)
Argentina	Inst. Naçional de Prevencion Sismica; Inst. Sismologico Zonda (2400 km)
New Zealand	Department of Scientific and Industrial Research (1370 km)
Fiji	Department of Mineral Resources (300 km)
Vanuatu Islands	ORSTROM, Cornell Univ. (620 km)
Papua New Guinea	Geological Survey (Port Moresby, Rabaul) (1700 km)
Taiwan	Central Weather Bur., Chinese Earthquake Res. Cent, Taipei (400 km)
Japan	Japan Meteorological Agency; Nat. Res. Inst. for Earth Science and Disaster Prevention; several universities on Honshu, Hokkaido, and Kyushu (2700 km)
Greece (Hellenic arc)	National Observatory of Athens (1040 km)

Information based on PRESGRAVE *et al.* (1985), and network descriptions in published research papers. The number in parentheses is the approximate linear extent of the network along the strike of the subduction boundary. Coverage may extend somewhat beyond the ends of the network. The density of stations within the networks is highly variable, as is the areal extent perpendicular to the subduction boundary.

subduction boundary is also shown. The density of coverage within this length is often not uniform, especially for the longer networks. The actual length of the subduction zone that is monitored by a network is somewhat greater than its length because events beyond the ends of the network can be detected and located, though with diminished completeness and accuracy. It should be noted that the two networks in the Aleutian Islands and the Lamont-Doherty network in the northeastern Caribbean have been terminated because of funding limitations. All of these were originally installed in the mid-1970s with support from the U.S. Geological Survey.

Earthquake Locations with Regional Network Data

One of the expected benefits from the use of densely spaced regional and local networks is that more accurate hypocenter locations can be calculated than can be accomplished with teleseismic data alone. This intuitive expectation, however, neglects the effects of the complicated velocity structure characteristic of subduction zones. Several authors have examined the relative accuracy of local network and teleseismic solutions, with the general conclusion that the local network solutions are better for shallow events (depths less than about 50–100 km) and the teleseismic solutions are better for intermediate-depth and deep events.

An early study of this problem, in which the value of regional network data is demonstrated, was conducted by UTSU (1975). His principal conclusions are supported by later work by others. He relocated 53 well-observed deep and intermediate-depth earthquakes in Japan with joint determination of lateral heterogeneity of velocity. He determined the P-wave residuals at each station from selected hypocenters and grouped the stations by their location relative to the volcanic front and the Japan trench. His premise was that residuals based on travel-times calculated without including lateral velocity variations in the upper mantle are related to the large-scale anomalies in the upper mantle resulting from the high-velocity oceanic lithosphere underthrusted beneath the island arc. Residuals from calculated travel-times for which such lateral heterogeneity has been included are controlled by crustal and upper mantle structure beneath each station. Areas of positive P-residuals derived with large-scale mantle heterogeneity included were found to be generally areas of positive Bouguer gravity anomalies.

UTSU (1975) describes the systematics of the mislocations of hypocenters computed by JMA and in the PDE catalog. For very deep events, the PDE locations seem more accurate, because the rays miss most of the anomalous velocity due to the underthrust slab. The JMA locations of very shallow events are judged to be more accurate because they are controlled by local stations and the paths are free of anomalous structure. To determine the "absolute" locations of deep and intermediate earthquakes with regional data, lateral heterogeneity of the upper mantle must be taken into account.

Additional studies of the problems of locating earthquakes in island arcs, and comparisons of teleseismic and local-network solutions has been carried out by ENGDAHL *et al.* (1977), BARAZANGI and ISACKS (1979), ENGDAHL *et al.* (1982), McLAREN and FROHLICH (1985), and NIEMAN *et al.* (1986). Special problems in using island-based local networks alone are discussed by LAWTON *et al.* (1982) and FROHLICH *et al.* (1982).

Barazangi and Isacks concluded, as UTSU (1975) had suggested, that the effect of descending slabs on teleseismic locations of mantle earthquakes (depth greater than 50 km) is smaller than the effect on locations based on local data. They judge that the incorporation of a descending slab in the velocity model is effective in the Aleutians, but not in Japan because the dip of the Wadati-Benioff (W-B) zone under Honshu is shallow, about 30°, and the aperture of the local network is wide. They conclude that the location of intermediate-depth earthquakes in the Aleutians based on teleseismic data are probably more representative of "absolute" locations than the locations based on local network data.

ENGDAHL *et al.* (1982) found that trench and intermediate-depth events in the central Aleutian Islands are biased little by the effects of slab structure on travel-times to teleseismic stations. Locations of intermediate-depth events based only on local data are systematically displaced oceanward (south), by an amount proportional to depth. Shallow events on the main thrust zone are located well by local data, but severely shifted northward and deeper, in the direction of the subduction, for teleseismic locations. They conclude that the velocity structure in the slab is not known well enough to permit accurate locations of thrust zone events without the use of local data.

McLAREN and FROHLICH (1985) presented a useful summary of location problems. They focused on the principal cause of these problems, the effect of the high-velocity subducted slab on travel-times. They used a simple model to explore the gross effects of the slab dip, the velocity contrast between the slab and the mantle, the depth of the earthquake inside the slab, as well as the effects of the gross geometry of the network. From their model results, they point out that the flat-earth assumption is satisfactory for networks smaller than about 2.5° and that P wave only solutions are inferior to those using both P and S waves. Conventional location methods applied in subduction zones may yield large location errors with small residuals, or accurate locations with large residuals. The residuals are not reduced by accurate arrival-time picks or by using both P and S waves. RMS residuals are not a good indicator of location quality in subduction settings.

The biggest problem revealed by the modeling is that deeper events are moved oceanward and deeper, as observed by ENGDAHL *et al.* (1982), so that the apparent configuration of the W-B zone incorrectly shows oversteepening or change in dip (Figure 1). The thickness of the W-B zone is in error by amounts that change with depth, and a double-planed zone in the input model apparently converges to a single plane at depth, giving a distribution similar to those that have been reported as real by others. Of the network configurations tested, linear distributions of seismometers parallel to the strike of the subducting slab with limited extent perpendicular to the slab yield the largest mislocations of the deeper events. This is a common configuration in island arcs.

Three-dimensional ray-tracing in velocity structure derived from thermal models was also employed by NIEMAN et al. (1986) to investigate further the problem of teleseismic mislocations. As found by others, the greatest mislocation errors relative to local network solutions are for events in the main thrust zone. They used a comparison of observed mislocations in the central Aleutian Islands with model results to constrain the key variables characterizing the models, the depth of penetration of the downgoing slab and the thermal coefficient of seismic velocity. They concluded that the downgoing slab in the central Aleutians extends at least 100 km deeper than the deepest observed seismicity.

LAWTON et al. (1982) and FROHLICH et al. (1982) report the results for studies of locations for which the land stations in island-based networks are combined with ocean-bottom seismographs (OBS). FROHLICH et al. (1982) evaluated the detection and location capability of the Central Aleutians Seismic Network from an experiment in which eight University of Texas OBS were operated for six weeks about 50 km south of Adak Island, over a very active segment of the seismic zone. A significant finding disclosed that events occur on the outer trench slope that are never seen by the island stations because they are in a shadow zone for this source area produced by the velocity structure.

Locations of hypocenters calculated from the island stations alone were compared with those for which the data from the OBS were included. The epicenters of events in the main thrust zones were almost identical, and most events detected by the OBS were seen by the island stations. The depths of these events from island-station observations alone were often in error, as will be discussed below. The epicenters of events deeper than 70 km, determined by the island station data alone, were 10 to 80 km south of the locations based on the combined data. This is one of the effects modeled by MCLAREN and FROHLICH (1985).

The accuracy of locations based only on local data and location using a flat-layered model is improved by the use of station corrections. Such corrections are essential if consistent solutions are to be obtained when data from part of the network are not available. Because of the three-dimensional variation in velocity structure that is typical of subduction zones, different corrections are required at each station for events in different parts of the region being monitored. Ray-tracing through a model for every small event to be located imposes a heavy computational burden. Empirical corrections may be determined as the mean residuals at a station from numerous well-recorded events in each defined subregion (ENGDAHL et al., 1982). Corrections based on ray-tracing in models of the velocity structure are effective and comparison of the computed station corrections with observed station residuals has been used to select the best-fitting model (HAUKSSON, 1985).

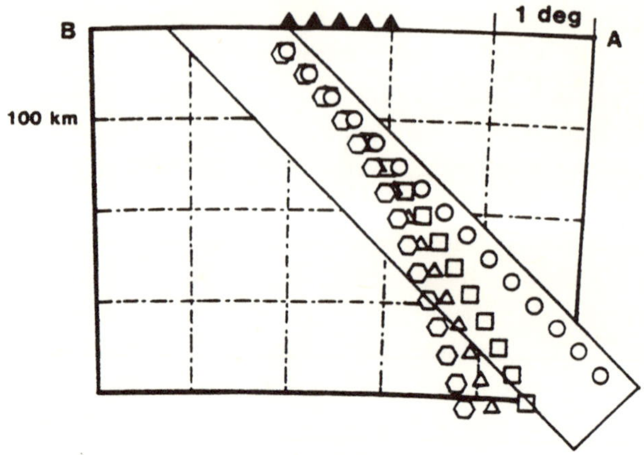

Figure 1

Modeled effects of the slab/mantle velocity contrast on earthquake locations, when the slab structure is neglected. The circles are the original positions of the hypocenters in the model. Travel-times to the stations were calculated by tracing rays through the simplified structure, after which the hypocenters were relocated from these times using a conventional location program. The relocations were based on a 4 percent slab/mantle contrast (squares), a 7 percent contrast (open triangles), and a 10 percent contrast (hexagons). From MCLAREN and FROHLICH (1985).

It is obvious that the depths determined for events far outside the network are subject to grave errors because of the trade-off between depth and origin time. In the case of the Central Aleutians Seismic Network, hypocenters tend to cluster in a depth range 20–26 km, with an apparent gap in activity down to about 30 km, Figure 2 (ENGDAHL et al., 1982, Figure 2; TOTH and KISSLINGER, 1984). This pattern is due to the effect of a boundary in the velocity model at a depth of 26 km and the unsatisfactory source-network geometry.

TOTH and KISSLINGER (1984) describe a useful approximate method for finding more accurate depths for events for which both P and S waves are recorded at a number of stations, and which are deep enough that an adequate number of stations receive upward traveling waves. The event origin time is determined without reference to a velocity model from the graph of (S-P) vs. P (the Wadati diagram). The depth and average velocity between the hypocenter and the surface are then calculated on the basis of a straight ray approximation (the $X^2–T^2$ method). The pattern of depth corrections found by TOTH and KISSLINGER (1984), using local data only, agrees closely with that of FROHLICH et al. (1982) for solutions using the OBS data (Figure 3). When locations of events were recalculated with the depth fixed at the value found by this method, it was found that the epicenters moved very little, in agreement with the results of FROHLICH et al. (1982).

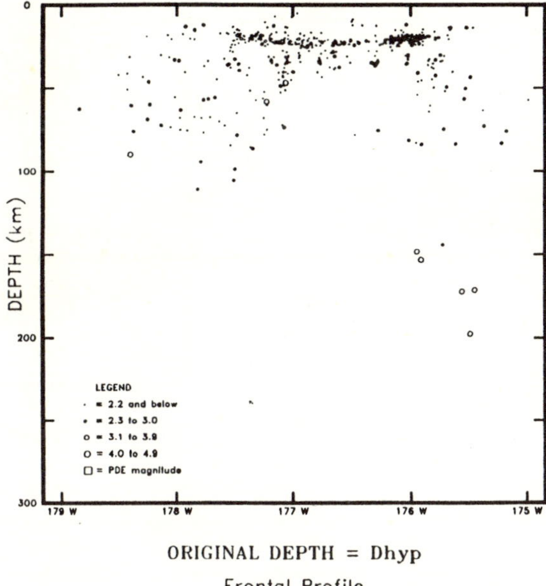

ORIGINAL DEPTH = Dhyp

Frontal Profile

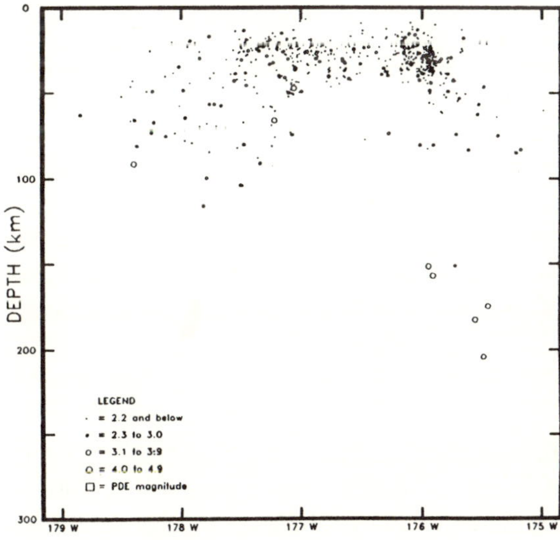

REVISED DEPTH = Dcorr

Frontal Profile

Figure 2

Vertical section parallel to the Aleutian arc of hypocenters located with data from the Central Aleutians Seismic Network. *Top*: the original depth distribution using the conventional location program and flat-layered velocity model; *bottom*: the revised depths from the X^2–T^2 analysis. The concentration of events near 25 km, with a gap to about 30 km, in the original solutions is an artifact due to a boundary in the velocity model at 26 km. This band of events becomes smoothly distributed with depth in the revised solutions. From TOTH and KISSLINGER (1984).

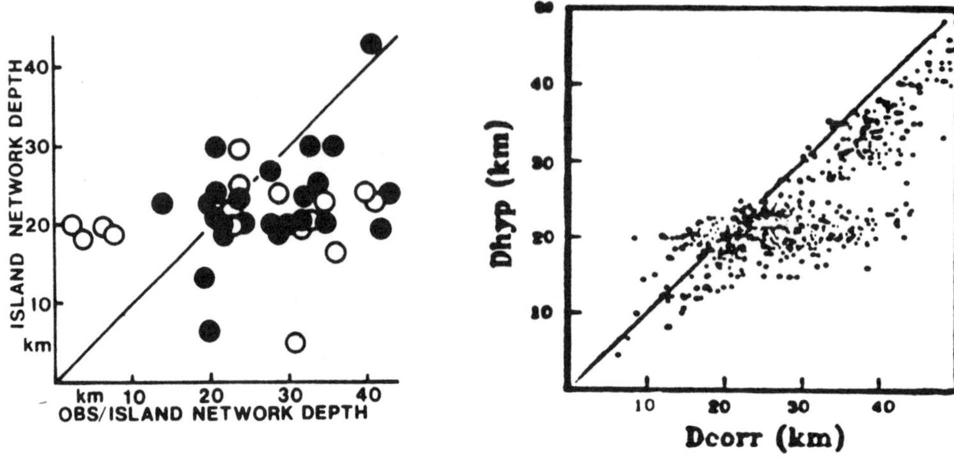

Figure 3

A comparison of the depth revisions when OBS data are combined with the island stations, left, and the revisions from the X^2-T^2 solutions, right. The data from the OBS/Island combination are from FROHLICH et al. (1982); the results on the right are from TOTH and KISSLINGER (1984). The open circles in the left panel mark poorer solutions.

Subduction-zone Configurations and Processes

Many studies of the distribution of earthquakes, the configuration of the W-B zone, and the velocity structure in subduction zones have been carried out on the basis of local and regional network data. Only a few of these are summarized here, with special emphasis on results obtained that would have been impossible with teleseismic data alone. Many of the authors invert the data for velocity structure by use of the method developed by AKI and LEE (1976) or a variation of that method.

North Pacific

Japan and its surrounding regions is the classical type area for detailed studies of subduction-zone seismicity based on data provided by high-quality regional and local seismic networks (UTSU, 1971). The large quantity and high quality of the data now available for research may be judged from the short summaries of observations by every seismological research organization in Japan, published since 1970, now twice each year, by the Geographical Survey Institute, Ministry of Construction, in the Report of the Coordinating Committee for Earthquake Prediction (Vol. 47 was published in February, 1992). Thousands of epicenters are mapped, along with vertical sections and focal mechanisms of interesting events. A comprehensive review of the many research projects that have used the Japanese data base is beyond the scope of this paper. Only two examples of subduction-zone configuration studies

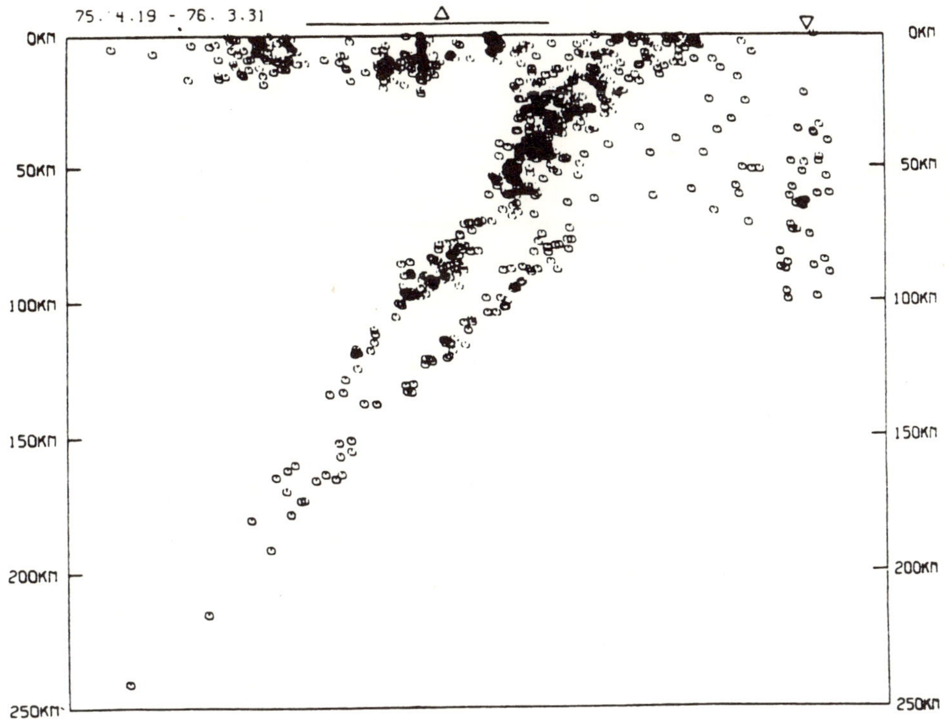

Figure 4

The double-planed Wadati-Benioff zone under northeastern Honshu, Japan. This is an east-west section of hypocenters between 39°N and 40°N, for events between April, 1975 and March, 1976. From HASEGAWA *et al.* (1978a).

that would have been impossible without the data from dense regional networks are given. These cited papers contain numerous references to earlier work.

Northeast Honshu. The mapping of the double-planed structure of the W-B zone under the northeastern Japan arc, Figure 4 (HASEGAWA *et al.* 1978a,b) is an outstanding example of a detailed subduction zone study based on the distribution of hypocenters carefully located with regional network data. The data were obtained with the Tohoku University seismic network which monitors all of northern Honshu Island. They found that the seismicity below about 50 km is concentrated in two parallel thin sheets, each about 10 km thick, separated by 30–40 km. They further concluded that the upper boundary of the descending slab lies in the upper plane of the double-plane deep seismic zone.

KAWAKATSU (1986) summarizes briefly the history of the discovery of double-planed W-B zones and lists places for which such configurations have been reported. He develops a model for the formation of double-planed zones on the basis of compatibility of seismic strain released at intermediate depths and strains due to unbending of the descending plate. He suggests constraints on the rheology of the plate imposed by the model results.

Kanto-Tokai Region. The Kanto-Tokai region of central Japan is a very complicated tectonic province (ISHIDA, 1992). Three plates (Philippine Sea, Eurasian, and Pacific) and two island arcs interact to produce high seismicity and geological complexity. Ishida summarizes earlier attempts by herself and others to portray the geometry of the plate configurations on the basis of hypocenter locations and presents a revised model. Her recent results are based on the relocation of 47,500 events recorded at four or more stations of the network operated by the National Research Institute for Earth Science and Disaster Prevention. The relocations were done with a revised three-dimensional velocity model, and the focal mechanisms of many selected events were also produced from this data set.

The distribution of hypocenters and high velocity zones was interpreted by ISHIDA (1992) to produce contour maps of the top of the Philippine Sea Plate and the Pacific Plate under the Kanto-Tokai region (Figure 5). The directions of the motion of these two plates relative to the Eurasian Plate were interpreted from the focal mechanisms as well as the orientation of the contours (northwestward for PHS-EUR, westward for PAC-EUR). She concluded that her new model provides a consistent explanation for the focal mechanisms of earthquakes larger than magnitude 6, and accounts for the localization of earthquake clusters under Kanto by the interaction at depths of the two subducted plates.

Eastern Aleutians. A comparable detailed study of the structure of the subduction zone in the eastern Aleutian arc was conducted by REYNERS and COLES (1982). They utilized data from the Lamont-Doherty Shumagin Islands network to map hypocenter distributions and to determine composite focal mechanisms. They found that a double-plane W-B zone exists there also, similar in configuration to that under northeastern Honshu. They proposed a model of the deformation processes active in this subduction zone. Subduction under northeastern Japan and the eastern Aleutians is comparable because of the relative simplicity of the subduction geometry.

HUDNUT and TABER (1987) offer evidence that the double-planed zone in the region monitored by the Shumagin Islands network is present only west of a boundary that crosses the main seismic zone at about 160.5°W longitude, trending normal to the arc at an azimuth at about 330°. To the east of this boundary, the W-B seismicity falls in a single planar zone. They infer a change in the stress state across this boundary. They suggest that perhaps the ratio of aseismic to seismic slip is greater on the east side (single-plane side) or that the two sides are at different stages of the strain accumulation cycle.

Southwest Pacific

Papua New Guinea. ABERS and ROECKER (1991) used *P* and *S* arrival times from the regional network in Papua New Guinea to relocate earthquakes in an

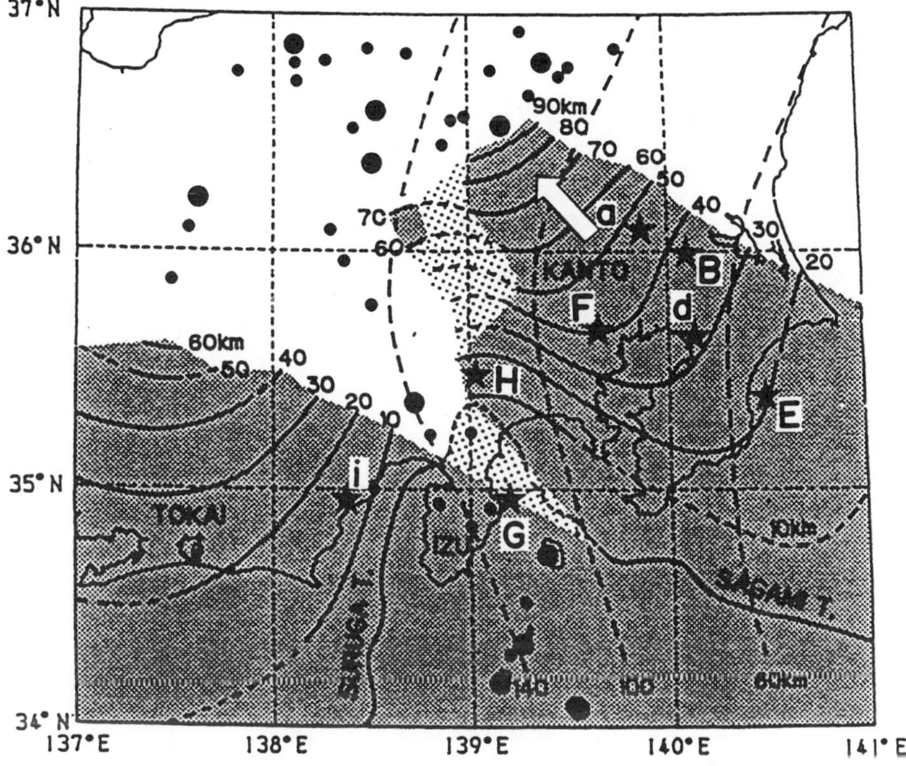

Figure 5

Contour map of the upper boundary of the Philippine Sea Plate (solid and short dashed lines) and the Pacific Plate (long dashed lines) under the Kanto region of Japan. The shaded area is the configuration of the Philippine Sea Plate, from distributions of hypocenters and high velocity zones. The white arrow shows the direction of suduction of the Philippine Sea Plate under the Eurasian Plate; the direction of subduction of the Pacific Plate is almost due west. The solid circles are Quaternary volcanoes. The letters mark earthquakes for which focal mechanisms are given in the original paper. From ISHIDA (1992).

arc-continent collision zone, with the objectives of clarifying subcrustal lithospheric deformation and the processes by which subduction ceases. They point out that the morphology of the deep seismic zone under New Guinea is poorly known because the standard teleseismic catalogs show "a diffuse cloud of hypocenters beneath northern New Guinea. Because many earthquake locations in these catalogs are of unknown and sometimes poor quality, it is difficult to be sure that the apparently wide scatter in locations is real and not a result of location errors."

Arrival times from almost 1000 well-recorded events were used in an inversion procedure that simultaneously located the hypocenters and produced a three-dimensional velocity structure. An example of the detailed results of ABERS and ROECKER (1991) in which the resolution possible with a regional network is compared with the teleseismic results is shown in Figure 6. The earthquake

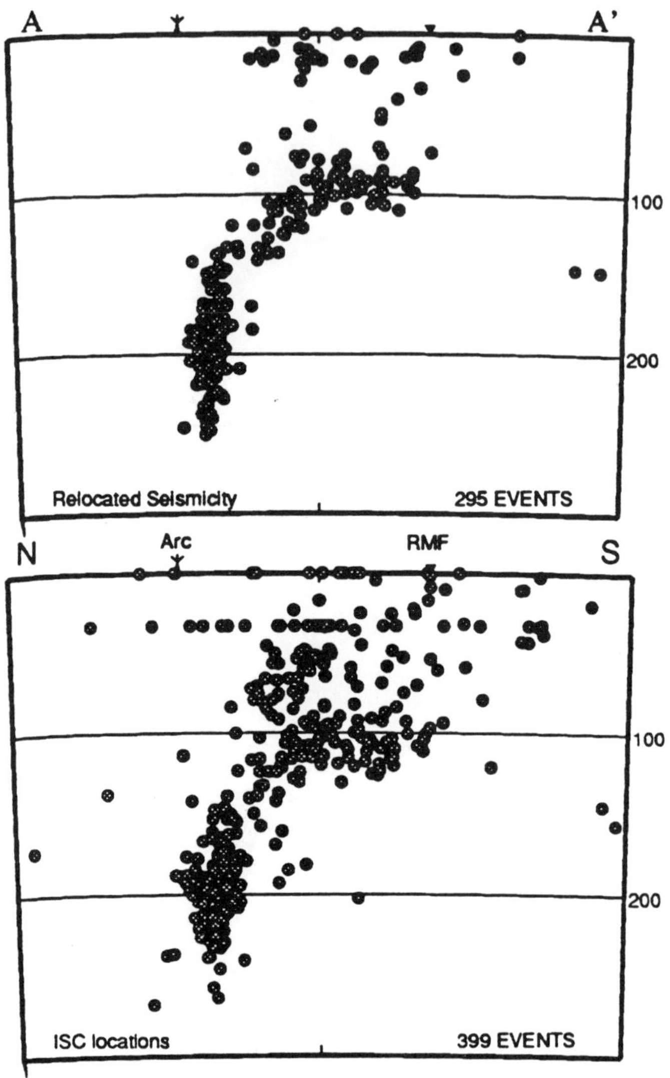

Figure 6
Vertical sections of hypocenters under eastern Papua New Guinea, in which locations relocated with regional data, top, can be compared with those from the ISC catalog, bottom. The vastly better definition of the configuration of the seismogenic zones is obvious. The position on the profile of the volcanic arc and the Ramu-Markham fault (RMF) is indicated. From ABERS and ROECKER (1991).

locations define a north dipping slab under northern New Guinea, flat at a depth of 100 km beneath the Huon-Finisterre ranges, and plunging almost vertically to 250 km beneath the Bismarck volcanic arc. This figure also illustrates the difference in the resolution of the configuration permitted by the regional network data and the ISC data.

Vanuatu Islands (New Hebrides Island Arc). The recent identification of a low *P*-velocity structure embedded in a prominent gap in seismicity within the W-B zone of the New Hebrides subduction zone is the latest in a sequence of studies of this region carried out by a team from Institut Français de Recherche Scientifique pour le Development en Cooperation (ORSTROM), Cornell University, and Rensselaer Polytechnic Institute (PREVOT *et al.*, 1991, and the accompanying references). The regional network operated by ORSTROM and Cornell since 1978 (mapped in ROECKER *et al.*, 1988) was supplemented for the recent study by eight OBS deployed by the University of Texas. *P* and *S* arrival times from a selected set of 4181 events were used to determine the three-dimensional structure. These data were supplemented by teleseismic arrival times of selected events reported by the ISC.

The authors infer that differences in the spatial distribution of seismicity seen in vertical sections normal to the arc at various latitudes are not due to large-scale changes in the structure of the descending slab, but to local anomalies within it. The coincidence of low *P* velocity in the slab and an absence of seismicity is interpreted as due to a thermal anomaly that lowers the strength of the subducted plate. This anomaly may be due to the effect of the subducted D'Entrecasteaux Fracture Zone. They also note that the position of the low-velocity zone coincides with a region of active volcanism.

Wave arrivals produced by multipathing and wave-type conversion have also been used to elucidate the details of subducted slab structure. One example, from the Vanuatu arc, is the analysis of multiple wave arrivals by CHIU *et al.* (1985). In their investigation of the effect of the slab on wave propagation, they identified two arrivals on short-period seismograms from earthquakes deeper than 70 km that were waves refracted in the slab at a *P*-to-*S* conversion at the presumed upper surface of the slab. They conclude that the intermediate-depth earthquakes occur within a zone of lower *P*-wave velocity above the colder inner core of the slab, and that the transition at the top of the slab to the surrounding upper mantle has a finite thickness, or the material above the slab attenuates high-frequency shear waves strongly.

Cascadia Subduction Zone

Comprehensive summaries of the seismicity of the Cascadia subduction zone, to which regional network data contribute substantially, are presented by ROGERS and HORNER (1991) (Canadian part) and LUDWIN *et al.* (1991) (Washington and Oregon). The Cascadia subduction zone has been the subject of detailed investigations of the geometry of a subducted slab for which the data provided by a dense regional network were essential (CROSSON and OWENS, 1987; WEAVER and BAKER, 1988). The first paper reported results under western Washington and southern British Columbia; the second extended the analysis to include northern Oregon. The configuration of the subducted Juan de Fuca Plate was taken as

defined by the distribution of carefully located subcrustal hypocenters and (CROSSON
and OWENS, 1987) the analysis of *Ps* converted phases from teleseismic events.

The finding in both studies is that the Juan de Fuca Plate is arched upward
under Puget Sound (Figure 7). The contours of the subducted oceanic Mohorovicic
discontinuity are constrained by the results of independent controlled-source studies
in places far from abundant seismicity. Subcrustal seismicity is concentrated along
the axis of the arch. The two papers present interpretations of the significance of the
arching of the subducted plate, with the accompanying change in the direction of

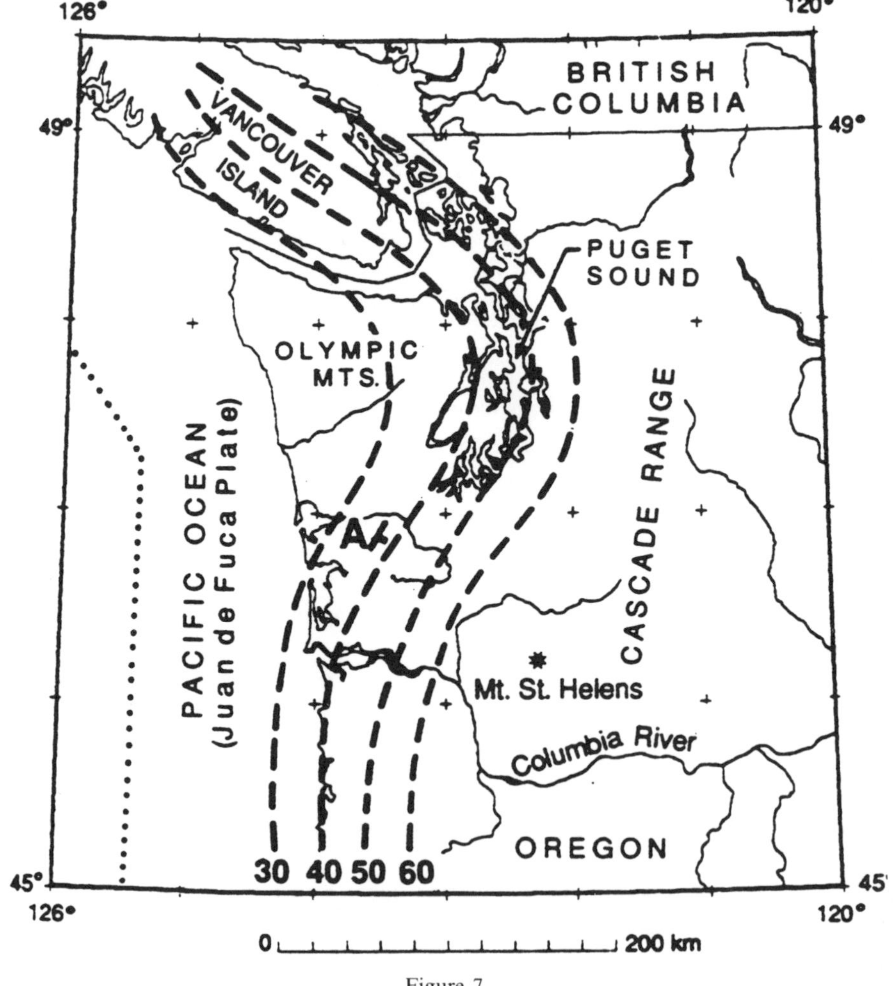

Figure 7

Proposed contour map (depths in km) of the subducted Mohorovicic discontinuity in the Cascadia
subduction zone. The dotted line is the estimated position of the trench. The arching of the subducted
Juan de Fuca Plate under Puget Sound is seen. From CROSSON and OWENS (1987).

the dip of the slab north and south of the axis, possible relations with surface topographic and geologic features, and a number of unresolved questions.

The Nazca Plate

Central and Southern Peru. As pointed out by HASEGAWA and SACKS (1981), western South America is an excellent site for study of a major subduction zone at which an oceanic slab descends under a continent. The circumstance provides an opportunity to investigate contrasts with island arc subduction.

A primary objective of their study was to resolve disagreement about the dip angle, whether about 10° or 30°, of the subducting Nazca Plate under central Peru and under southern Peru and northern Chile. As they point out, ISC and PDE hypocenter locations are poor for this region, and "local seismic network data are essential to resolve this kind of controversy." They used data from nine stations of the local Arequipa network, which covered part of southern Peru for which no disagreement existed, and central Peru, for which differing models had been proposed.

Their findings, from carefully relocated hypocenters (Figure 8) show that the subducting plate dips at an angle of about 30° on all vertical sections crossing the study area, to a depth of about 100 km, bends almost horizontally at that depth in the northern part of the region, but continues with a constant dip in the southern part. They conclude that the data show distortion of a continuous slab, not a tear in the descending plate.

Western Argentina. A detailed study of the configuration of the subducted Nazca Plate under northwestern Argentina was done by SMALLEY and ISACKS (1987). They emphasize the importance of knowing the thickness, morphology, and internal structure of W-B zones. After summarizing the problems of hypocenter locations based on local network and teleseismic data, discussed above, they point out that an ideal arrangement is to have data from sites directly above a nearly horizontal Wadati-Benioff zone, thereby avoiding the effects of a dipping slab. Their research area is one of the few places at which observations can be made of rays from deep sources traveling almost vertically through a flat structure.

They used P arrivals from the regional network (see Table 1) to fix the epicenters, P and S times to fix the depths, and S-P time intervals to obtain the depth and thickness of the W-B zone. They conclude that the observations fit a single-planed W-B zone and that the hypocenters are intraplate, in the subducted slab, not interplate.

The Caribbean

The Caribbean Sea Plate and its interactions with the North American, South American, Nazca, and Cocos Plates present some of the most intriguing and

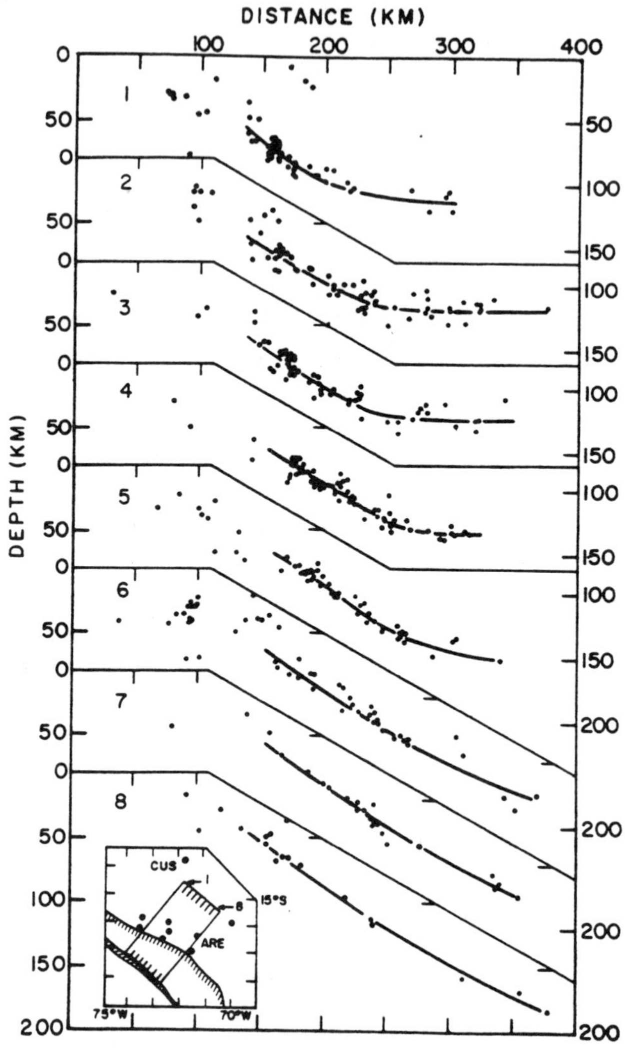

Figure 8

Vertical sections of relocated hypocenters, Peru, along profiles indicated in the inset. The tectonically important differences in the configuration below 100 km of the Wadati-Benioff zone from northwest to southeast is unambiguous. From HASEGAWA and SACKS (1981).

challenging problems of contemporary tectonics research. Detailed seismic monitoring of the subduction zones along the eastern and northeastern boundaries of the plate began only in 1973 with the French network in the Lesser Antilles (DOREL, 1981) and in 1975 with the U.S.G.S. network in Puerto Rico and the Lamont-Doherty Geological Observatory (LDGO) network, which spanned from Puerto Rico to the northern Lesser Antilles (MURPHY and McCANN, 1979).

DOREL (1981) combined the data from the French network with a catalog of historical felt earthquakes. He found that the seismicity tends to cluster in the northern half of the Lesser Antilles arc. He estimated the rate of subduction of the North American Plate under the Caribbean Plate as 2 cm/yr. Hypocenter distributions along six profiles normal to the arc show a well-defined W-B zone, with dip steepening from about 30° at the north end to about 60° in the center, and apparently becoming less steep again to the south. Hypocenters near the center of the arc are as deep as 190 km.

FRANKEL *et al.* (1980) mapped the hypocenter distribution in the Virgin Islands segment of the northeastern boundary of the Caribbean Plate from the LDGO network data. In contrast to the Lesser Antilles arc, the relative plate motion is highly oblique to this portion of the boundary. Hypocenters deeper than 25 km align in a south-dipping configuration that suggests the subduction of the North American Plate. A large number of shallow swarms was also reported.

FRANKEL (1982) used LDGO network data to find composite focal mechanisms for microearthquakes in the same segment of the plate boundary. The focal mechanisms and the distribution of hypocenters suggest an oblique underthrusting of the North American Plate along this boundary.

b-values, Quiescence, and Aftershocks

The large number of events located by regional networks make possible studies of earthquake distributions down to microearthquake magnitudes. The nonlinearity of the frequency-magnitude distribution is a fundamental problem of current interest (PACHECO *et al.*, 1992). TAYLOR *et al.* (1990) examined the distribution of earthquakes under and near southeastern Hokkaido, Japan. Working from a data base of over 11,000 earthquakes located with the Hokkaido University network within 50 km of the Carnegie Institution station, KMU, during 1976–87, they identified two "effectively decoupled suites" of events: crustal (Eurasian Plate) and subduction (Pacific Plate). Events in the main thrust zone were put in the "crustal" category.

Aftershocks of the Urakawa-oki earthquake (March 21, 1982, $M_{JMA} = 7.1$) and another strong event on January 14, 1987 were eliminated by restricting the time intervals retained to 1977–1981 and 1983–1986. They tested the remaining 3306 crustal earthquakes and 743 subduction events for completeness, using a test based on time-of-day of occurrence, devised by RYDELEK and SACKS (1989). Although both data sets were evaluated as complete to duration magnitude, m_d, 2.0, the frequency-magnitude distribution of the crustal events departed from linearity below $m_d = 2.5$, and the subduction events below $m_d = 3.5$. The logarithm of the ratio of the number of crustal events to the number of subduction events was found to depend linearly on magnitude from $m_d = 0$ to $m_d = 5$. From this, the authors conclude that the nonlinearity is a real effect, with a common cause for the two suites. They did not suggest the cause.

The Urakawa-oki earthquake was also the study of possible precursory quiescence (TAYLOR *et al.*, 1991). From the data set of 35,000 earthquakes located during July, 1976 to December, 1986, they concluded that a significant quiescence existed for events with magnitude greater than 3 during two years before the main shock, but activity at magnitudes less than 2.4 increased near the main shock epicenter. The minimum magnitude for detection and location by the local network was estimated at 1.0–1.5. One of the lessons from this study is that in the search for seismicity rate changes that may be related to the preparation for an imminently strong earthquake, it is necessary to have a large enough data set so that tectonically interesting small divisions of the overall seismic zone can be examined separately, with enough events in them to yield statistically meaningful results. This can only be done for typical seismogenic zones from catalogs of microearthquakes.

A case of seismic quiescence that was observed and reported as a prediction prior to the occurrence of a major event was related to the May 7, 1986 Andreanof Islands earthquake, $M_W = 8.0$ (KISSLINGER, 1988). Earthquake counts for $m_d = 2.3$ and above, from May, 1976 through December, 1985, revealed a strong and persistent quiescence in part of the main thrust zone, starting at different times in parts of the monitored region, but most clearly in mid-1982. The quiescence was most prominent in what became the area of greatest moment release during the main event (HWANG and KANAMORI, 1986; DAS and KOSTROV, 1990; HOUSTON and ENGDAHL, 1989).

Although the decrease in activity is also seen in the teleseismically located seismicity (ENGDAHL *et al.*, 1989), the number of these events was too small to permit a statistically acceptable verification of the behavior seen in the microearthquake data. Since the seismic zone was initially divided into strips 20 km wide, east-to-west, it is obvious that only the data from a local network could have provided a basis for the analysis. This attempt at the prediction of a major earthquake is evaluated in KISSLINGER (1988).

Teleseismic catalogs show very few aftershocks of intermediate and deep earthquakes. Only with the aid of a regional network capable of detecting and locating microearthquakes within a subducted plate can such aftershock sequences be adequately documented. The potential for gaining new information regarding earthquake processes at intermediate depths from the analysis of locally recorded aftershock sequences is suggested by the unexpected behavior exhibited in the two examples analyzed by KISSLINGER and HASEGAWA (1991). Both sequences, one in the central Aleutians (Figure 9) the other under northeastern Honshu, Japan, included a late surge of activity similar to the secondary aftershocks found for crustal earthquakes, but without the occurrence of a strong aftershock to trigger them. The cause is unknown, but may be related to creep in the thrust zone above the hypocenters or in the subducted plate. The decay characteristics shown in Figure 9 are very similar to those of crustal sequences.

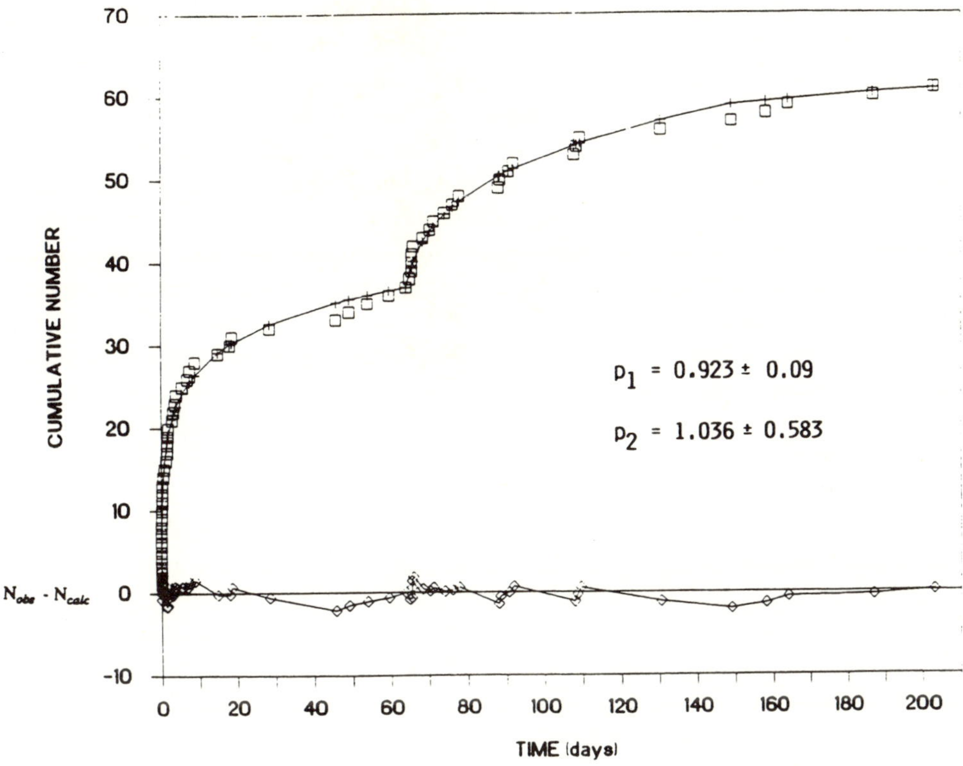

Figure 9

The temporal behavior of an intermediate-depth aftershock sequence (mainshock $m_b = 6.0$; depth = 105 km) in the central Aleutian Islands. The data were modeled with the modified Omori relation. A secondary sequence, which was not triggered by a strong aftershock, is apparent. The Omori p values for the two parts of the sequence are shown, and are very similar to those for crustal earthquakes. A minimum duration magnitude of 1.9 was used. The misfit of the model is shown at the bottom. From KISSLINGER and HASEGAWA (1991).

Coda-Q Observations

Changes in the physical properties of highly strained rocks are expected to occur during the time of preparation of a volume for a large earthquake. Although the search for reliable precursors of earthquakes on the basis of such changes has not yet been successful, the testing of a number of hypotheses continues. One suggested precursor is a change in attenuation of seismic waves induced in rocks near failure (AKI, 1985). The waveforms recorded by local stations are required for the analysis and interpretation of coda-Q as an indicator of stress-induced changes in the attenuative properties of the lithosphere.

Stress changes can also cause small changes in the elastic wave velocity in materials and introduce velocity anisotropy. Such changes can result in small time-dependent mislocations of earthquakes when a fixed velocity model is used. This effect can be detected and accounted for by the analysis of earthquake doublets (POUPINET *et al.*, 1984). A doublet is a pair of earthquakes that occur in essentially the same place, perhaps widely separated in time, and produce almost identical waveforms at all stations. Cross-spectral analysis of the two seismograms at each station yields highly accurate relative locations of the two events (POUPINET *et al.*, 1984) and can be used to separate differences in location from changes in material properties (KAZI, 1990).

Some examples of investigations of the frequency dependence of coda-Q and its variations with space and time in subduction zones are those by RODRÍGUEZ *et al.* (1983), GUSEV and LEMZIKOV (1985), SCHERBAUM and KISSLINGER (1985), and HUANG and KISSLINGER (1992). All these authors employed the technique of AKI and CHOUET (1975), or some modification of it.

RODRÍGUEZ *et al.* (1983) and SCHERBAUM and KISSLINGER (1985) were interested primarily in the frequency dependence of coda-Q. The first group used seismograms from two local digital stations installed near Petatlan by the University of Wisconsin. They found a rather strong frequency dependence of coda-Q, as $f^{0.87}$, and somewhat higher attenuation at Petatlan than at Oaxaca. SCHERBAUM and KISSLINGER (1985) worked from seismograms of the Central Aleutians Seismic Network and found that coda-Q varied as $f^{1.03}$.

GUSEV and LEMZIKOV (1985) and HUANG and KISSLINGER (1992) were interested in the possible spatial variation of coda-Q, as well as temporal changes that might be associated with the preparation for and occurrence of a strong earthquake. The former authors first worked to separate the effects of anelastic absorption from those due to scattering, and concluded that the effects were comparable in a frequency range of 0.4–6 Hz, with absorption predominant at higher frequencies. Evaluation of temporal changes in coda shape were based on data from the regional network of Kamchatka-Kurile Islands. A parameter based on differences at each time between the actual waveform amplitude and the amplitude predicted by a standard coda-decay function was used as the criterion for the coda shape. Anomalies significant at the 95% level were found preceding three major earthquakes in the region. Gusev and Lemzikov argue that the most probable cause in the change in the coda shape is change in absorption, not in heterogeneity. They also identify features in the amplitude-frequency spectrum that correlate closely with the times of anomalous coda shape. They ascribe these to localized scatterers that may be concentrated very near the source area of the impending strong earthquake.

HUANG and KISSLINGER (1992) were looking for possible precursory changes in coda-Q associated with the 1986 M_W 8.0 Andreanof Islands earthquake. Various combinations of sources in the main thrust zone and groups of stations of the

Central Aleutians Seismic Network were used to look first for any spatial distribution of coda-Q prior to the strong event. It was found that coda-Q in the western part of the 1986 rupture zone (the eastern part of the rupture was outside of the area covered by the network) was about 15% higher than in the part of the thrust zone to the west that did not rupture. No significant variations with time prior to the main event could be identified. Data from aftershocks showed a decrease of coda-Q of about 10% during approximately two months after the main shock.

Identification of Asperities

The identification of asperities and barriers on the main thrust zone is one goal of subduction-zone seismology. Strong sections of faults in the thrust zone may act as asperities at which rupture nucleates or as sites of large slip during major earthquakes. They may be barriers that stop rupture that nucleated elsewhere on the fault. Moment release patterns derived from the analysis of teleseismic recordings of large earthquakes serve well to define the distribution of slip after the fact. Some experiments have been performed in an attempt to define asperities prior to rupture. Two such studies have been carried out with data for a small seismogenic zone from the Central Aleutians Seismic Network (Figure 10), one based on clustering of moderate magnitude earthquakes, $m_b \geq 4.5$ (BOWMAN and KISSLINGER, 1985), the other on patterns of stress drops and apparent stresses of microearthquakes in the same place (SCHERBAUM and KISSLINGER, 1984).

BOWMAN and KISSLINGER (1985) became interested in this site because of activation in the form of six moderate earthquakes between July, 1980 and April, 1983, following an interval of six years with only two teleseismically recorded events. The local-network locations of all PDE-reported events within the subregion were clustered in two tight groups. The b-value for events in this subregion was found to be the lowest of any subdivision of the main thrust zone, 0.7 compared to 0.97 for the whole thrust zone. The clustering and the low b-value (high mean magnitude) suggested that this site is an area of stress concentration, possibly marking an asperity on the main thrust zone. It proved to be within the western extremity of the rupture zone of the 1986 great earthquake (KISSLINGER, 1988). SCHERBAUM and KISSLINGER (1984) found a clustering of relatively high stress-drop events in the same small subregion and a significant increase in apparent stresses and stress drops during a three-year interval prior to $m_b = 5.8$ earthquake in one of the cluster's moderate events.

Work on the details of the mechanical properties of the seismically active structures and of the rupture process are research areas that would be greatly enhanced by the wider distribution of broadband, three-component instrumentation in subduction zones.

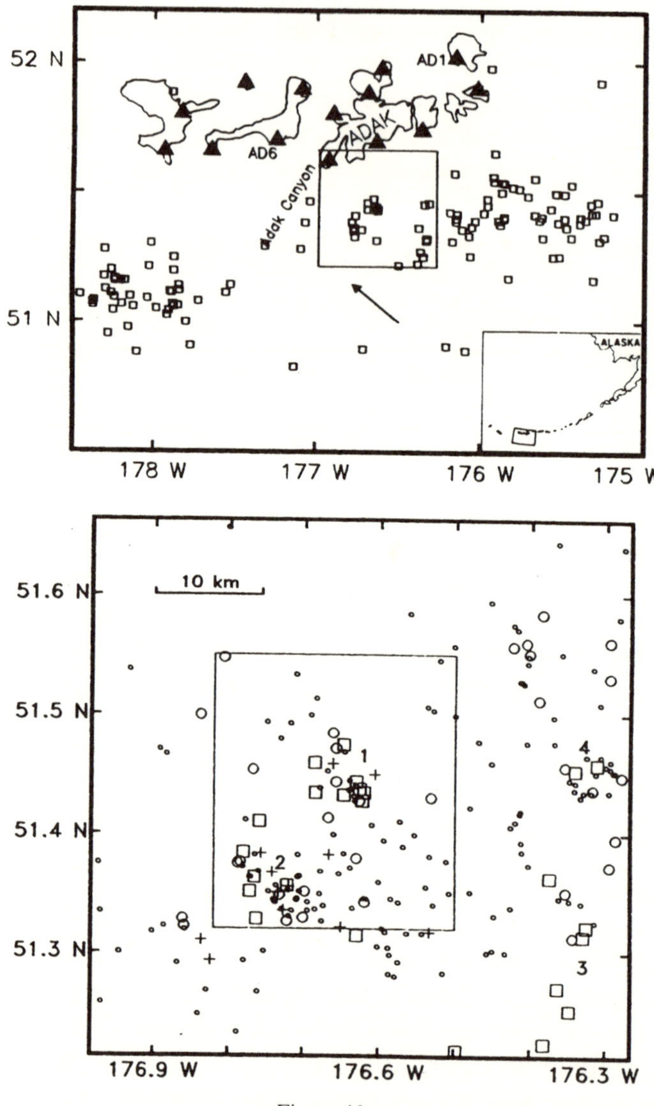

Figure 10

Top: Map of the Central Aleutians Seismic Network (triangles) and local network solutions of all events reported by the U.S. Geological Survey for August 1974 through June, 1983 (squares). The rectangle outlines the area shown in the bottom part of this figure. *Bottom*: Map of local network solutions of events with depths less than 70 km, $m_d \geq 2.5$, for August, 1974 through June, 1983. Squares mark events strong enough to be in the PDE catalog; circles are smaller events. The suspected asperity is within the rectangle. The clustering of stronger events is seen. From BOWMAN and KISSLINGER (1985).

Summary and Discussion

The unique and essential contributions of local and regional networks as data sources for subduction-zone research are demonstrated by the results of many recent studies, a sample of which has been reviewed here. Earthquakes down to the microearthquake level are detected and located. Occasional events as small as magnitude 1 may be located but local catalogs seem to be complete in most cases only for events greater than about magnitude 2.5. A somewhat higher magnitude cutoff may be needed for completeness of intermediate or deep earthquakes. On the other hand, standard catalogs based on teleseismic observations are reasonably complete only for events with M_b greater than about 4.5–5, depending on the location of the subduction zone. The problems of obtaining accurate relative hypocenter locations are understood and are manageable with careful analysis. In particular, the locations of numerous events in the main thrust zone from local data are more reliable than those achievable with teleseismic data alone.

The large number of well-located events in the best local and regional catalogs have enabled detailed studies of both the spatial and temporal distribution of the seismicity in individual subduction zones. The spatial distribution of seismicity may be interpreted in terms of the configuration of the interface between the colliding plates and the subducted slab. This distribution is also necessary for associating seismic activity with specific geologic or topographic features. The cases selected for brief review here illustrate the potential of such studies for clarifying the details of the kinematics of the subduction process in a variety of regional settings. Differences in the mode of subduction beneath island arcs and continents have been brought out by these studies. Earthquake distributions by themselves cannot fix the boundaries of subducting plates, although investigators make assumptions about the relation of the hypocenter positions to the plate surfaces. Analysis of multiple wave arrivals from intermediate-depth events has aided in defining the position of the upper plate surface and the velocity distribution within the plate. More work on this problem is likely in the future.

The multitude of events in regional and local catalogs is essential for investigations of the temporal behavior of seismicity in small segments of subduction zones. Detection of significant variations in background seismicity and determination of the spatial distribution and decay properties of aftershocks are greatly enhanced by regional data complete to small magnitudes. Studies of aftershocks of intermediate-depth events are especially difficult without regional data because so few aftershocks of such earthquakes are found in standard teleseismic catalogs.

Many regional networks report magnitudes calculated from coda durations. These magnitudes may be consistent for events processed for a single network, but a study should be made of the consistency of duration magnitudes reported by different networks. Network operators should also document any changes made in the method of calculating magnitudes, as these may introduce systematic magnitude shifts that adversely affect statistical analyses of the network catalog (HABERMANN and CRAIG, 1988).

The application of local network data to studies of source characteristics and properties of the rocks within the active zones has been limited by the small dynamic range and narrow passband of most network seismographs, as well as the sparsity of three-component recording. Replacement of the present instrumentation with equipment that overcomes these limitations will probably happen slowly, one network at a time. Great improvement in subduction-zone seismicity studies would result also from the more extensive use of ocean-bottom seismographs to overcome the limitations on instrument distribution imposed by the locations of land areas suitable as station sites. Both locations and focal mechanism solutions will be improved by better azimuthal distribution of observations.

Data from closely spaced three-component broad-band stations will contribute to the determination of the physical properties of the materials within the subduction system. More detailed delineation of the distribution of seismic wave velocities, velocity anisotropy, and anelastic attenuation is needed to constrain the thermal state and the properties of flow in the subducting plate and the surrounding upper mantle.

Future research based on local and regional network data will most likely continue to develop more details of the distribution of seismicity within parts of subduction zones that are monitored well. It is expected that with more data and careful analysis, more detailed three-dimensional velocity and attenuation models will be obtained. With more and better hypocenter locations, the relation of earthquake occurrence to regional tectonics and specific geological structures can be better evaluated. The distribution of inhomogeneities of strength and other properties of the fault zone materials that control earthquake occurrence is another goal for the future. Advances in understanding source processes on the basis of local data require the use of instruments with higher dynamic range than is currently employed, so that moderate to large events are recorded without clipping. The calculation of focal mechanisms of small earthquakes from P-wave first motion polarities is difficult for linear networks deployed parallel to the strike of the subduction boundary, because of the limited azimuthal coverage. Mechanisms derived by matching broad-band waveforms at even a few stations will add considerably to the data base available for tectonic interpretations.

Additional studies of the configuration and ambient physical conditions in and around the subducted plate, the main thrust zone, and associated regional features will eventually yield a global picture of the highly variable properties of subduction zones and insight into the factors controlling these properties. A good start has been made toward this goal.

Regional data are also important for earthquake hazard assessment in subduction zones, as they are in all seismically active places. The identification of the active structures on the basis of the distribution of small earthquakes is especially important in subduction zones, where the most significant faults are often inaccessible for direct observation. The general level of activity in different parts of the zone can be determined more rapidly and in finer detail than is possible with teleseismic data alone. Although the attempts to detect precursors to great earthquakes by seismic observations have met with limited success, the data provided by the frequent small earthquakes may contain evidence diagnostic of the preparation for great earthquakes that is still to be discovered and understood.

Acknowledgements

The installation and operation of the Central Aleutians Seismic Network and the research in the Central Aleutian Islands subduction zone described in this paper has been supported since 1974 by a Cooperative Agreement and a series of grants from the U.S. Geological Survey. The aftershock study mentioned was partially supported by a grant from the National Science Foundation. Appreciation is expressed to the staff scientists, data analyst, and technicians in Boulder, St. Louis, and on Adak who made this project possible.

REFERENCES

ABERS, G. A., and ROECKER, S. W. (1991), *Deep Structure of an Arc-Continent Collision: Earthquake Relocation and Inversion for Upper Mantle P and S Wave Velocities beneath Papua New Guinea*, J. Geophys. Res. *96*, 6379–6401.

AKI, K. (1985), *Theory of Earthquake Prediction with Special Reference to Monitoring of the Quality Factor of Lithosphere by the Coda Method*, Earthq. Predict. Res. *3*, 219–230.

AKI, K., and CHOUET, B. (1975), *Origin of Coda Waves: Attenuation and Scattering Effects*, J. Geophys. Res. *80*, 3322–3342.

AKI, K., and LEE, W. H. K. (1976), *Determination of Three-dimensional Velocity Anomalies under a Seismic Array Using First P Arrivals from Local Earthquakes. I. A Homogeneous Initial Model*, J. Geophys. Res. *81*, 4381–4399.

BARAZANGI, M., and ISACKS, B. (1979), *Comparison of the Spatial Distribution of Mantle Earthquakes Determined from Data Produced by Local and by Teleseismic Networks for the Japan and Aleutian Arcs*, Bull. Seismol. Soc. Am. *69*, 1763–1770.

BOWMAN, J. R., and KISSLINGER, C. (1985), *Seismicity Associated with a Cluster of Earthquakes of $m_b \geq 4.5$ near Adak, Alaska: Evidence for an Asperity?*, Bull. Seismol. Soc. Am. *75*, 223–236.

CASTRO, R. R., ANDERSON, J. C., and SINGH, S. K. (1990), *Site Response, Attenuation and Source Spectra of S Waves along the Guerrero, Mexico, Subduction Zone*, Bull. Seismol. Soc. Am. *80*, 1481–1503.

CHIU, J. M., ISACKS, B. L., and CARDWELL, R. K. (1985), *Propagation of High-frequency Seismic Waves Inside the Subducted Lithosphere from Intermediate-depth Earthquakes Recorded in the Vanuatu Arc*, J. Geophys. Res. *90*, 12,741–12,754.

CHIU, J. M., STEINER, G., SMALLEY, Jr., R., and JOHNSTON, A. C. (1991), *PANDA: A Simple, Portable Seismic Array for Local- to Regional-scale Seismic Experiments*, Bull. Seismol. Soc. Am. *81*, 1000–1014.

CROSSON, R. S., and OWENS, T. J. (1987), *Slab Geometry of the Cascadia Subduction Zone beneath Washington from Earthquake Hypocenters and Teleseismic Converted Waves*, Geophys. Res. Letters *14*, 824–827.

DAS, S., and KOSTROV, B. V. (1990), *Inversion for Seismic Slip Rate History and Distribution with Stabilizing Constraints: Application to the 1986 Andreanof Islands Earthquake*, J. Geophys. Res. *95*, 6899–6913.

DOREL, J. (1981), *Seismicity and Seismic Gap in the Lesser Antilles Arc and Earthquake Hazard in Guadeloupe*, Geophys. J. Roy. Astr. Soc. *67*, 679–695.

ENGDAHL, E. R., SLEEP, N. H., and LIN, M. T. (1977), *Plate Effects in North Pacific Subduction Zones*, Tectonophys. *37*, 95–116.

ENGDAHL, E. R., DEWEY, J. W., and FUJITA, K. (1982), *Earthquake Location in Island Arcs*, Phys. Earth and Planet. Int. *30*, 145–156.

ENGDAHL, E. R., BILLINGTON, S., and KISSLINGER, C. (1989), *Teleseismically Recorded Seismicity before and after the May 7, 1986, Andreanof Islands, Alaska, Earthquake*, J. Geophys. Res. *94*, 15,481–15,498.

FRANKEL, A. (1982), *A Composite Focal Mechanism for Microearthquakes along the Northeastern Border of the Caribbean Plate*, Geophys. Res. Letters *9*, 511–514.

FRANKEL, A., MCCANN, W. R., and MURPHY, A. J. (1980), *Observations from a Seismic Network in the Virgin Islands Region: Tectonic Structures and Earthquake Swarms*, J. Geophys. Res. *85*, 2669–2678.

FROHLICH, C., BILLINGTON, S., ENGDAHL, E. R., and MALAHOFF, A. (1982), *Detection and Location of Earthquakes in the Central Aleutian Subduction Zone using Island and Ocean Bottom Seismograph Stations*, J. Geophys. Res. *87*, 6853–6864.

GUSEV, A. A., and LEMZIKOV, V. K. (1985), *Properties of Scattered Elastic Waves in the Lithosphere of Kamchatka: Parameters and Temporal Variations*, Tectonophys. *112*, 137–153.

HABERMANN, R. E., and CRAIG, M. S. (1988), *Comparison of Berkeley and CALNET Magnitude Estimates as a Means of Evaluating Temporal Consistency of Magnitudes in California*, Bull. Seismol. Soc. Am. *78*, 1255–1267.

HASEGAWA, A., UMINO, N., and TAKAGI, A. (1978a), *Double-planed Structure of the Deep Seismic Zone in the Northeastern Japan Arc*, Tectonophys. *47*, 43–58.

HASEGAWA, A., UMINO, N., and TAKAGI, A. (1978b), *Double-planed Deep Seismic Zone and Upper-mantle Structure in the Northeastern Japan Arc*, Geophys. J. Roy. Astr. Soc. *54*, 281–296.

HASEGAWA, A., and SACKS, I. S. (1981), *Subduction of the Nazca Plate beneath Peru as Determined from Seismic Observations*, J. Geophys. Res. *86*, 4971–4980.

HAUKSSON, E. (1985), *Structure of the Benioff Zone beneath the Shumagin Islands, Alaska: Relocation of Local Earthquakes Using Three-dimensional Ray Tracing*, J. Geophys. Res. *90*, 635–649.

HOUSTON, H., and ENGDAHL, E. R. (1989), *A Comparison of the Spatio-temporal Distribution of Moment Release for the 1986 Andreanof Islands Earthquake with Relocated Seismicity*, Geophys. Res. Lett. *16*, 1421–1424.

HUANG, Z. X., and KISSLINGER, C. (1992), *Coda-Q before and after the 1986 Andreanof Islands Earthquake*, Pure and Appl. Geophys. *138*, 1–16.

HUDNUT, K. W., and TABER, J. J. (1987), *Transition from Double to Single Wadati-Benioff Seismic Zone in the Shumagin Islands, Alaska*, Geophys. Res. Lett. *14*, 143–146.

HWANG, L. J., and KANAMORI, H. (1986), *Source Parameters of the May 7, 1986 Andreanof Islands Earthquake*, Geophys. Res. Lett. *13*, 1426–1429.

ISHIDA, M. (1992), *Geometry and Relative Motion of the Philippine Sea Plate and Pacific Plate beneath the Kanto-Tokai District, Japan*, J. Geophys. Res. *97*, 489–513.

KAWAKATSU, H. (1986), *Double Seismic Zones: Kinematics*, J. Geophys. Res. *91*, 4811–4825.

KAZI, W. M. (1990), *Velocity Changes in the Central Aleutian Islands Determined from the Analysis of Earthquake Doublets Using Cross Spectral Analysis Method*, Ph.D. Dissertation, University of Colorado.

KISSLINGER, C. (1988), *An Experiment in Earthquake Prediction and the May 7, 1986 Andreanof Islands Earthquake*, Bull. Seismol. Soc. Am. *78*, 218–229.

KISSLINGER, C., and HASEGAWA, A. (1991), *Seismotectonics of Intermediate-depth Earthquakes from Properties of Aftershock Sequences*, Tectonophys. *197*, 27–40.

LAWTON, J., FROHLICH, C., PULPAN, H., and LATHAM, G. V. (1982), *Earthquake Activity at the Kodiak Continental Shelf, Alaska, Determined by Land and Ocean Bottom Seismograph Networks*, Bull. Seismol. Soc. Am. *72*, 207–220.

LUDWIN, R. S., WEAVER, C. S., and CROSSON, R. S., *Seismicity of Washington and Oregon. In Neotectonics of North America* (Slemmons, D. B., Engdahl, E. R., Zoback, M. D., and Blackwell, D. D. eds.) (Geological Society of America 1991) pp. 77–98.

MCLAREN, J. P., and FROHLICH, C. (1985), *Model Calculations of Regional Network Locations for Earthquakes in Subduction Zones*, Bull. Seismol. Soc. Am. *75*, 397–413.

MURPHY, A. J., and MCCANN, W. R. (1979), *Preliminary Results from a New Seismic Network in the Northeastern Caribbean*, Bull. Seismol. Soc. Am. *69*, 1497–1513.

NIEMAN, T. L., FUJITA, K., and ROGERS, Jr., W. J. (1986), *Teleseismic Mislocation of Earthquakes in Island Arcs — Theoretical Results*, J. Phys. Earth *34*, 43–70.

PACHECO, J. F., SCHOLZ, C. H., and SYKES, L. R. (1992), *Changes in Frequency-size Relationship from Small to Large Earthquakes*, Nature *355*, 71–73.

POUPINET, G., ELLSWORTH, W. L., and FRÉCHET, J. (1984), *Monitoring Velocity Variations in the Crust Using Earthquake Doublets: An Application to the Calavaras Fault, California*, J. Geophys. Res. *89*, 5719–5731.

PRESGRAVE, B. W., NEEDHAM, R. E., and MINSCH, J. H. (1985), *Seismograph Station Codes and Coordinates, 1985 Edition*, U.S. Geological Survey Open-File Report 85–714.

PREVOT, R., ROECKER, S. W., ISACKS, B. L., and CHATELAIN, J. L. (1991), *Mapping of Low P Wave Velocity Structures in the Subducting Plate of the Central New Hebrides, Southwest Pacific*, J. Geophys. Res. *96*, 19,825–19,842.

PUJOL, J., CHIU, J. M., SMALLEY, R., REGNIER, M., ISACKS, B., CHATELAIN, J.-L., VLASITY, J., VLASITY, D., CASTANO, J., and PUEBLA, N. (1991), *Lateral Velocity Variations in the Andean Foreland in Argentina Determined by with the JHD Method*, Bull. Seismol. Soc. Am. *81*, 2441–2457.

REYNERS, M., and COLES, K. S. (1982), *Fine Structure of the Dipping Seismic Zone and Subduction Mechanics in the Shumagin Islands, Alaska*, J. Geophys. Res. *87*, 356–366.

RODRÍGUEZ, M., HAVSKOV, J., and SINGH, S. K. (1983), *Q from Coda Waves near Petatlan, Guerrero, Mexico*, Bull. Seismol. Soc. Am. *73*, 321–326.

ROECKER, S. W., CHATELAIN, J.-L., ISACKS, B. L., and PREVOT, R. (1988), *Anomalously Deep Earthquakes beneath the New Hebrides Trench*, Bull. Seismol. Soc. Am. *78*, 1984–2007.

ROGERS, G. C., and HORNER, R. B., *An overview of western Canadian seismicity*. In *Neotectonics of North America* (Slemmons, D. B., Engdahl, E. R., Zoback, M. D., and Blackwell, D. D. eds.) (Geological Society of America 1991) pp. 69–76.

RYDELEK, P. A., and SACKS, I. S. (1989), *A Test for Completeness of Earthquake Catalogs*, Nature *337*, 251–254.

SCHERBAUM, F., and KISSLINGER, C. (1985), *Coda Q in the Adak Seismic Zone*, Bull. Seismol. Soc. Am. *75*, 615–620. (1987), *Erratum*, Bull. Seismol. Soc. Am. *77*, 2246.

SCHERBAUM, F., and KISSLINGER, C. (1984), *Variations of Apparent Stresses and Stress Drops Prior to the Earthquake of 6 May 1984 ($m_b = 5.8$) in the Adak Seismic Zone*, Bull. Seismol. Soc. Am. *74*, 2577–2592.

SEISMOLOGY AND VOLCANOLOGY RESEARCH DIVISION, MRI (1980), *Permanent Ocean-bottom Seismograph Observation System*, Tech. Reports of the Meteorol. Res. Inst. 14.

SMALLEY, Jr., R. F., and ISACKS, B. L. (1987), *A High-resolution Local Network Study of the Nazca Plate Wadati-Benioff Zone under Western Argentina*, J. Geophys. Res. *92*, 13,903–13,912.

TAYLOR, D. W. A., SNOKE, J. A., SACKS, I. S., and TAKANAMI, T. (1980), *Nonlinear Frequency-magnitude Relationships for the Hokkaido Corner, Japan*, Bull. Seismol. Soc. Am. *80*, 340–353.

TAYLOR, D. W. A., SNOKE, J. A., SACKS, I. S., and TAKANAMI, T. (1991), *Seismic Quiescence before the Urakawa-oki Earthquake*, Bull. Seismol. Soc. Am. *81*, 1255–1271.

TOTH, T., and KISSLINGER, C. (1984), *Revised Focal Depths and Velocity Model for Local Earthquakes in the Adak Seismic Zone*, Bull. Seismol. Soc. Am. *74*, 1349–1360.

UTSU, T. (1971), *Seismological Evidence for Anomalous Structure of Island Arcs with Special Reference to the Japanese Region*, Rev. Geophys. and Space Phys. *9*, 839–890.

UTSU, T. (1975), *Regional Variations of Travel-time Residuals of P Waves from Nearby Deep Earthquakes in Japan and Vicinity*, J. Phys. Earth *23*, 367–380.

WEAVER, C. S., and BAKER, G. E. (1988), *Geometry of the Juan de Fuca Plate beneath Washington and Northern Oregon from Seismicity*, Bull. Seismol. Soc. Am. *78*, 264–275.

(Received May 6, 1992, revised September 30, 1992, accepted October 20, 1992)

PAGEOPH, Vol. 140, No. 2 (1993)

0033–4553/93/020287–14$1.50 + 0.20/0

Aftershocks of the San Marcos Earthquake of April 25, 1989 ($M_S = 6.9$) and Some Implications for the Acapulco–San Marcos, Mexico, Seismic Potential

F. R. Zúñiga,[1] C. Gutiérrez,[2] E. Nava,[3] J. Lermo,[3]
M. Rodríguez[3] and R. Coyoli[3]

Abstract—Aftershock activity following the April 25, 1989 ($M_S = 6.9$) earthquake near San Marcos, Guerrero, Mexico, was monitored by a temporary network installed twelve hours after the mainshock and remaining in operation for one week. Of the 350 events recorded by this temporary array, 103 were selected for further analysis in order to determine spatial characteristics of the aftershock activity. An aftershock area of approximately 780 km² is delimited by the best quality locations. The area of highest aftershock density lies inside an area delimited by the aftershocks of the latest large event in the region in 1957 ($M_S = 7.5$) and it partially overlaps the zone of maximum intensity of the earlier 1907 ($M_S = 7.7$) shock. Aftershocks also appear to cluster close to the mainshock hypocenter. This clustering agrees with the zone of maximum slip during the mainshock, as previously determined from strong motion records. A low angle Benioff zone is defined by the aftershock hypocenters with a slight tendency for the slab to follow a subhorizontal trajectory after a 110 km distance from the trench axis, a feature which has been observed in the neighboring Guerrero Gap. A composite focal mechanism for events close to the mainshock which also coincides with the zone of largest aftershock density, indicates a thrust fault similar to the mainshock fault plane solution.

The San Marcos event took place in an area which could be considered as a mature seismic gap. Due to the manner in which strain release has been observed to previously occur, the occurrence of a major event, overlapping both the neighboring Guerrero Gap and the San Marcos Gap segments of the Mexican thrust, cannot be overlooked.

Key words: Aftershocks, Guerrero, seismic gap, Mexico, subduction.

Introduction

The Acapulco–San Marcos (ASM) segment of the Mexican subduction zone ($99.0° – 100.0°$ W, Figure 1) has already surpassed the 30-yr criterion of McCann *et al.* (1979) for a mature seismic gap since the occurrence of a large ($M_S = 7.5$) earthquake in 1957. Thus it could be cataloged as such a seismic gap. The two

---

[1] Instituto de Geofísica, UNAM, Cd. Universitaria, México 04510, D.F.
[2] Centro Nacional de Prevención de Desastres, Delfín Madrigal 665, México 04360, D.F.
[3] Instituto de Ingeniería, UNAM, Cd. Universitaria, México 04510, D.F.

regions adjacent to the Acapulco–San Marcos segment are considered to have the highest potential for a large earthquake in the Mexican subduction zone, namely the Central Guerrero (CG) (approx. 100° to 101.2° W) (e.g., SUÁREZ *et al.*, 1990) and Ometepec (OM) (approx. 98.25° to 99.0° W) (e.g., GONZÁLEZ-RUIZ and McNALLY, 1988) seismic gaps (Figure 1). All three of these segments can be differentiated from the remaining Mexican thrust, in terms of source complexity, recurrence intervals, and multiplicity of events, making detailed studies of every earthquake occurrence an even higher priority if one is to elucidate possible mechanisms of stress accumulation and release in the region.

Some researchers (SINGH and MORTERA, 1991; SUÁREZ *et al.*, 1990; NISHENKO and SINGH, 1987) have suggested the possibility that all of the region from $101.2°$ to 99.1° W could be broken in a single magnitude $M_W > 8.0$ event. There is, of course, the alternative that strain energy might be released in a series of smaller ruptures as has occurred in the past during the sequence that took place from 1899 to 1911 (Figures 1 and 2).

Furthermore, seismic quiescence, at the teleseismic level, has also been identified for the area spanning from Acapulco to Ometepec starting in 1977 (McNALLY and MINSTER, 1981) as well as for the Ometepec segment (GONZÁLEZ-RUIZ and McNALLY, 1988) for the periods mid-1975 to early 1978, and early 1980 to mid-1982. It is unclear, however, whether the seismic rate decreases observed are inherent to natural causes or stem from changes in the reporting and/or detection of earthquakes in the area (HABERMANN, 1988). In the adjacent Guerrero gap, seismic quiescence has also been identified for a small central area (PONCE *et al.*, 1992) resembling a doughnut pattern.

A magnitude $M_S = 6.9$, earthquake took place on April 25, 1989, in the ASM region causing the death of at least three people in Mexico City. The epicenter of this event was very near the relocated epicenter of the 1957 shock (NISHENKO and SINGH, 1987; GONZÁLEZ-RUIZ and McNALLY, 1988). Portable array based locations of the aftershocks of the April 25, 1989 earthquake provide useful information that can be used to further understand the manner in which larger ruptures occur in the region. An aftershock analysis, based on data from a temporary local network, is presented supporting this purpose.

Previous Large Events in the Acapulco–San Marcos Region

Earlier studies on the ASM region (SINGH *et al.*, 1982; NISHENKO and SINGH, 1987) have proposed that the recurrence time for this segment of the Mexican subduction zone can be as long as 50 years, based on seismic moment estimates for the 1957 and the earlier 1907 ($M_S = 7.7$) event and on the idea that the 1957 shock may have been part of a delayed sequence of ruptures. NISHENKO and SINGH (1987) argue that the 1957 event appears to have broken approximately half of the rupture of the 1907 earthquake with the other half of the rupture occurring previously during earthquakes that occurred in 1937 and 1950 ($M_S = 7.5$, 7.1 respectively). They also assign a rupture length of ≈ 160 km for the 1907 mainshock. However, while they favor a case of delayed rerupture for the same segment, GONZÁLEZ-RUIZ and MCNALLY (1988) consider that both the 1937 and 1950 events broke a separate segment east of the rupture area of the 1957 shock (the Ometepec segment). They also assumed that the 1907 rupture, whose length they estimate as ≈ 110 km, did not encompass the volume ruptured by the 1937 and 1950 shocks. Thus, they propose that the OM segment ruptures are independent from those of the neighboring ASM region (Figure 1).

Recurrence intervals are particularly difficult to assess due to uncertainties in location and extent of rupture for past earthquakes. Recent attempts to relocate the

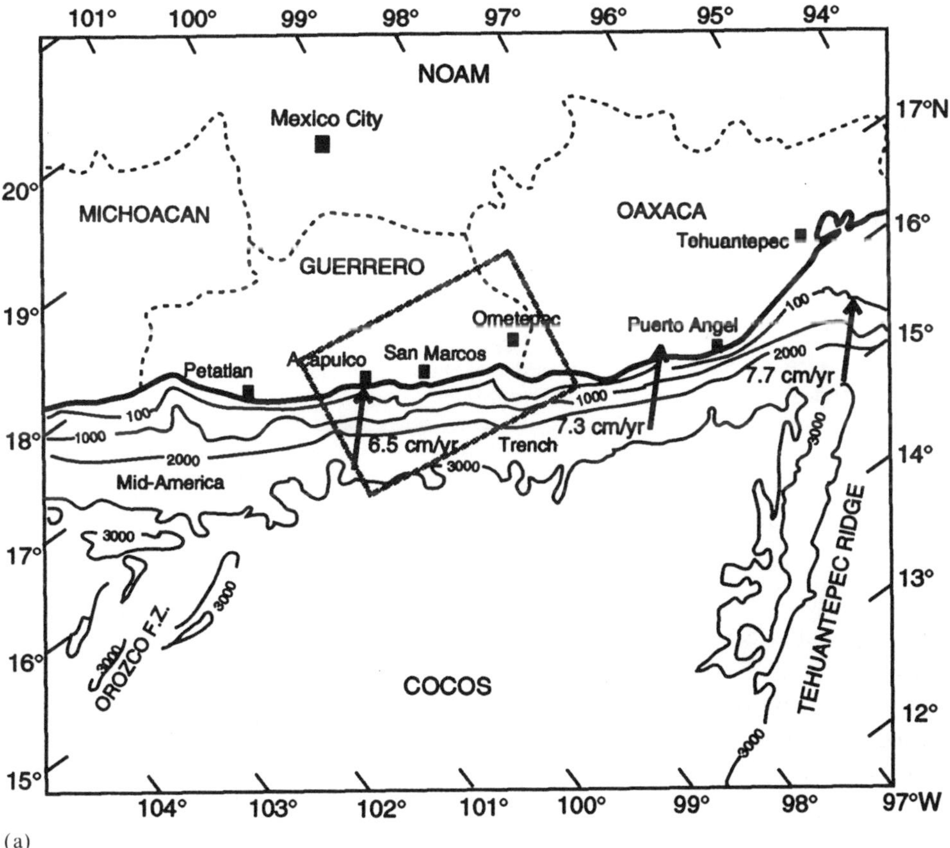

(a)

Figure 1(a)

Location map and tectonic situation of the San Marcos, Guerrero, Mexico, area. Bathymetry from CHASE *et al.* (1970). Solid arrows indicate the direction of relative convergence between the North American (NOAM) and Cocos or Rivera (RIV) plates (convergence vectors after MCNALLY and MINSTER (1981) and EISSLER and MCNALLY (1984)).

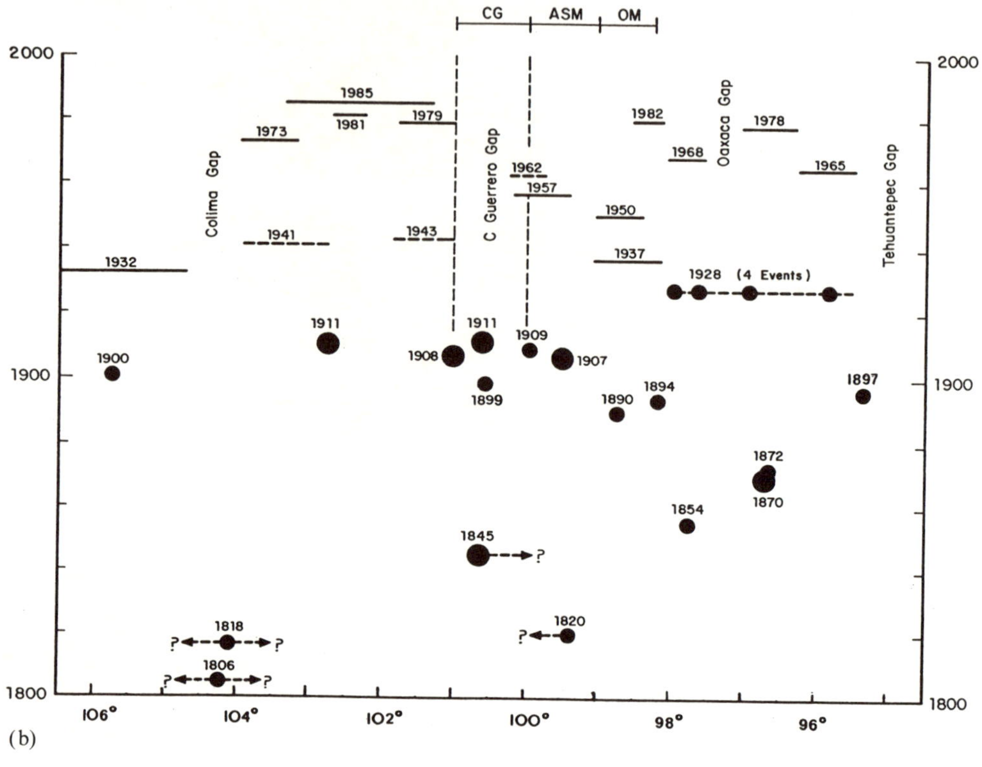

Figure 1(b)

Space-time diagram of historic and recent earthquakes ($M > 7.0$) along the Mexican subduction zone. Dashed horizontal lines with question marks represent unconstrained locations for events prior to 1850. Horizontal lines indicate the lateral extent of rupture zones, based on aftershock studies, these are dashed when in doubt. Modified after PONCE *et al.* (1992). The approximate extent of the Central Guerrero (CG), Acapulco–San Marcos (ASM) and Ometepec (OM) segments of the subduction zone are also indicated.

1957 aftershocks, by means of the Joint Hypocenter Determination (JHD) method (NISHENKO and SINGH, 1987; GONZÁLEZ-RUIZ and MCNALLY, 1988) have provided improved but differing locations. Thus, the rupture extent is still unclear due to the scarcity and poor quality of data available. In this study, we take González-Ruiz and McNally's estimate of the 1957 aftershock area as the most reliable since it is based on a two-step relocation which we consider provides a better resolution.

Historically, a great event that may have previously ruptured the ASM segment took place in 1845 with a reported magnitude of $M_S \approx 8.1$ (SINGH *et al.*, 1981) (Figures 1 and 2) and there is a possibility that an event that occurred in 1820 ($M_S \approx 7.6$) might have also ruptured the same volume (SINGH *et al.*, 1982). Figure 2 also shows other shocks whose location has been reported in the area

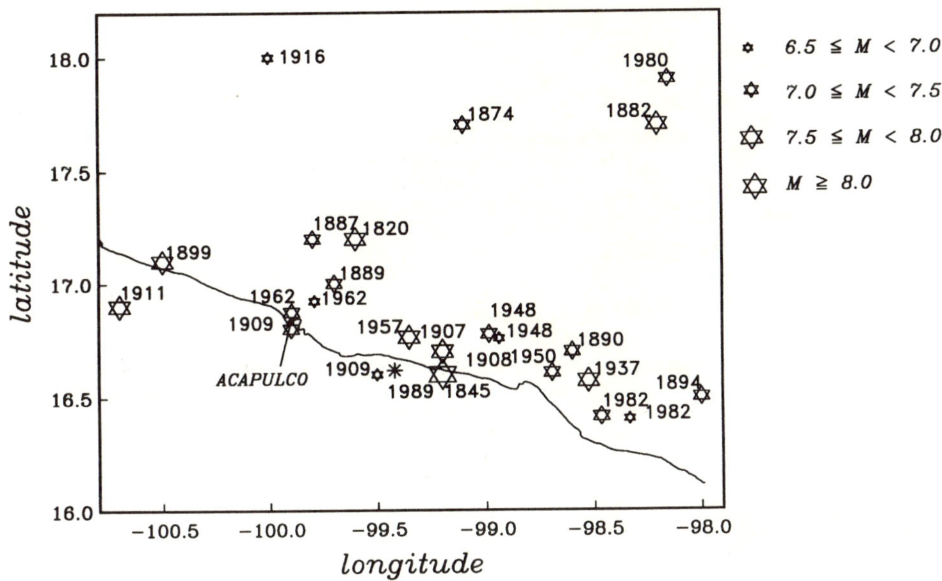

Figure 2

Epicenters of important earthquakes in the San Marcos vicinity. Location of 1937, 1948, 1950, 1957, 1962, 1980 and 1982 events after GONZÁLEZ-RUIZ and MCNALLY (1988) relocation study. All other events after ANDERSON *et al.* (1989) except for the 1989 mainshock (asterisk) whose location is from this study. The location of events in 1908 and 1909, shown here, are still uncertain.

(ANDERSON *et al.*, 1989), such as the 1908 ($M_S = 7.6$) and 1909 ($M_S = 6.9$) events. Due to uncertainties in location of these earthquakes, however, it is not clear whether they in fact occurred in the ASM region or further to the west (SINGH, pers. communication, 1992).

Aftershock Data

Twelve hours after the occurrence of the April 25, 1989 mainshock the first station of a temporary seismograph network started operating in the epicentral area and remained in operation for 7 days. The instruments deployed included six digital, three component, EDA/PRS-4 portable seismographs as well as six smoked paper Sprengnether/MEQ-800 portable seismographs belonging to the Institutes of Geophysics and Engineering of the National University of Mexico (UNAM). Sensors used were of the Kinemetrics 1 Hz Ranger type in the case of the analog stations while Lennartz LE-3D 1 Hz sensors were employed for the digital stations. A GOES receiver clock was used for the timing of digital data. Data was recorded on magnetic disks in both raw and processed formats with a sampling rate of 50 sps.

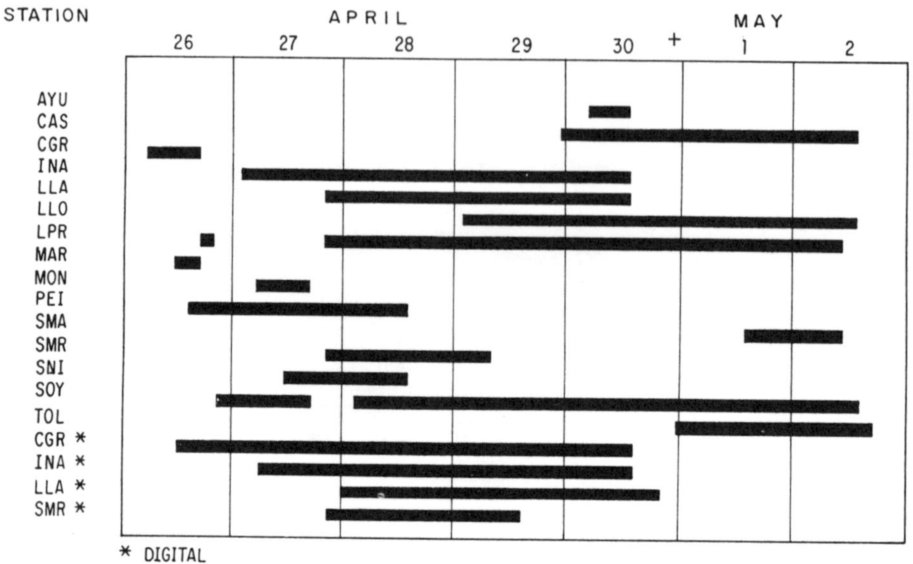

Figure 3
Operation schedule of stations of the temporary array, asterisks show digital stations. Solid lines indicate time when station operated.

The distribution of stations varied in order to improve coverage of the after-shock area as it was progressively defined, but the inaccessibility of some places prevented a better azimuthal coverage. The operation schedule is shown in Figure 3, station locations are shown in Figure 4. Some locations were shared by both kinds of instruments (digital and analogic) to provide a check for the operation of the event-triggered digital equipment and adjust the settings for triggering.

During the operation period, we registered more than 350 events of which 103 were selected for further analysis, based on the number of stations (4 minimum) that recorded them and the quality of first arrival picks. Hypocentral locations were determined using HYPO71 (LEE and LAHR, 1978) and the velocity model shown in Table 1. This model was recently obtained for the CG area (SUÁREZ *et al.*, 1992) by means of the Minimum Apparent Velocity method (MATUMOTO *et al.*, 1977). Both, P arrival times and S-P times were used in the location procedure. In order to better constrain the locations, the hypocenter of the largest aftershock ($M_S = 4.9$) was used as a master-event. Hypocentral location errors were acceptable if they were less than 3 km horizontally and less than 8 km vertically. Mean location RMS error was 0.22, the mean horizontal error was 3.24 km and the mean vertical error was 4.19 km. The after-shock epicenters distribution is shown in Figure 4, together with other reference features.

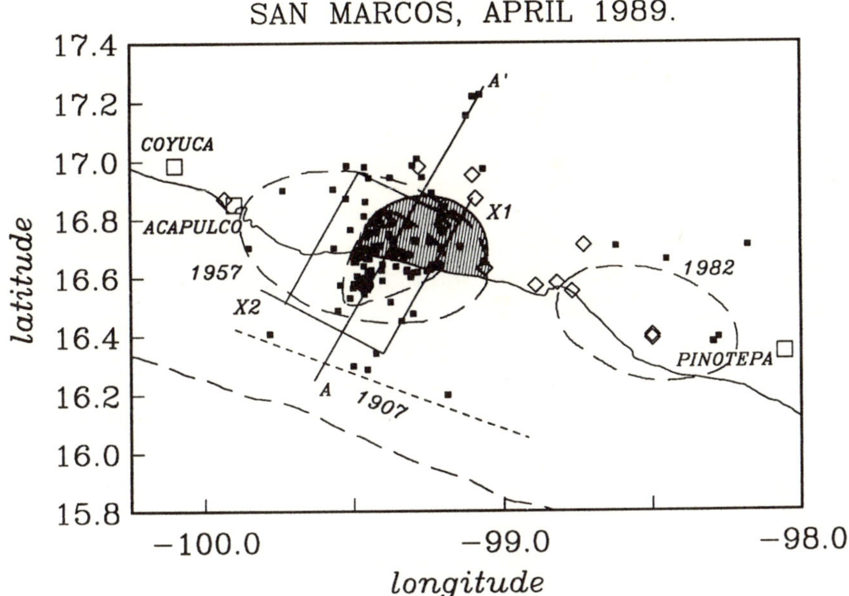

Figure 4

Aftershock epicenters for the week April 27–May 02, 1991. Also shown: stations (diamonds), main towns (squares) and projection lines (continuous straight lines) discussed in the text. A long-dashed line indicates the trench axis. A short-dashed line the rupture extent of the 1907 event, following GONZÁLEZ-RUIZ and MCNALLY (1988). The larger dashed ellipsoid is the aftershock area of the 1957 mainshock and the smaller ellipsoid is the corresponding aftershock area of the 1982 event, following GONZÁLEZ-RUIZ and MCNALLY (1988). The hatched semi-circular area indicates the maximum intensity zone for the 1907 event after BÖSE et al. (1908) and GONZÁLEZ-RUIZ and MCNALLY (1988). The aftershock area of the April 25, 1989 event is also outlined. A box used as reference by SINGH et al. (1989) in the inversion of strong motion data is indicated with axes X1 and X2 labeled. The location of the mainshock and largest aftershock is shown by large and small filled circles, respectively.

Table 1

Velocity model used in the location procedure. The model was determined using apparent velocities (SUÁREZ et al., 1992) of events in the Central Guerrero Gap area

Vel. (km/sec)	Depth (km)
5.8	0.0
6.5	10.14
7.1	18.36
7.4	23.42
8.0	34.00

Results and Discussion

The aftershock zone shown in Figure 4 was outlined using the best quality epicenter locations (63 in total) and considering the region with the highest density of events, giving an area of approximately 780 km². Figure 4 also shows the location of the mainshock (16.579° N, 99.462° W), determined from strong motion accelerograph records from the Guerrero Accelerograph Array (GAA) (ANDERSON *et al.*, 1986) together with the location of the largest aftershock (16.677° N, 99.483° W) for which it was possible to use records from GAA plus the temporary array. In the same figure we have also plotted the aftershock areas of the most recent events in the zone, i.e., 1957 and 1982 following GONZÁLEZ-RUIZ and MCNALLY (1988) coupled with the maximum intensity zone of the 1907 event (BÖSE *et al.*, 1908) and the proposed extent of its rupture after GONZÁLEZ-RUIZ and MCNALLY (1988).

As seen in Figure 4, the location of the 1989 mainshock and its aftershock zone overlap the aftershock zone for the 1957 mainshock. It is also noticeable that the zone of maximum intensity of the 1907 event, assumed to be the epicentral area, also overlaps both aftershock zones. Thus, this small area appears to stand out among the rest of ASM in terms of different stress release characteristics.

In the daily epicenter snap-shots shown in Figure 5 it can also be seen that aftershock activity is more tightly clustered and shows higher magnitudes near the south-western edge of the aftershock area (e.g., epicenter plots for April 28 and May 02 in Figure 5), very close to the location of the mainshock. A preliminary source study of the mainshock (SINGH *et al.*, 1989) based on the inversion of strong motion records from stations belonging to the GAA, some of which were located inside the aftershock area, showed that most of the slip appeared to be concentrated near the reported hypocentral location of the mainshock as well. Figure 6 (after SINGH *et al.*, 1989) shows results of that analysis for two different cases. The first corresponds to an unconstrained solution and the second to a solution constraining the slip in the $X1$ direction (for reference, the same box used in Figure 6 has been plotted in Figure 4, with axes $X1$ and $X2$ labeled in both figures). The shaded area in Figure 6b, is assumed to be the area that most contributed to the observed ground motion. These results are in apparent contradiction to results found for other events (e.g., Loma Prieta earthquake, WALD *et al.*, 1991; and Whittier Narrows earthquake, HARTZELL and IIDA, 1990), where the zone of maximum moment release appears to lack aftershocks.

Figure 7 shows a vertical profile using line A-A' (Figure 4) as reference. SUÁREZ *et al.* (1990) postulated the possibility that in the CG area the slab follows a subhorizontal trajectory after a distance of approximately 150 km from the trench. Focal mechanisms of small events in that area analyzed by them appear to confirm the idea of upward bending of the slab at this distance. A slight tendency to follow a subhorizontal trajectory in ASM can be discerned from Figure 7, starting after a

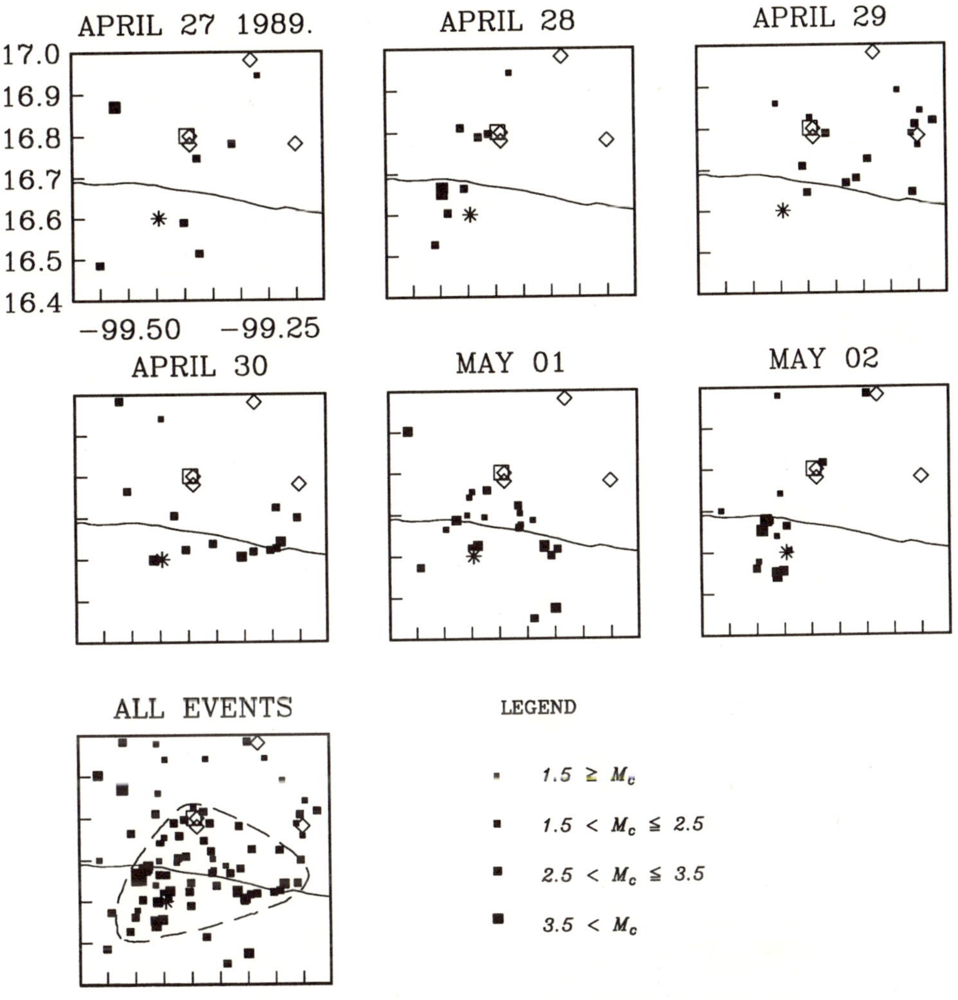

Figure 5

Daily epicenter maps for the period the array remained in operation. M_C stands for duration magnitude. The asterisk shows the location of the mainshock for reference. The square indicates the town of San Marcos. All other symbols as in Figure 4.

distance of about 110 km from the trench. At present, it is not known if this feature continues further east, however, hypocenters of events studied by PONCE *et al.* (1992) do not suggest that the subhorizontal trajectory continues near the Isthmus of Tehuantepec (at approximately 97.0° W).

A composite focal mechanism (Figure 8) obtained from aftershocks that occurred near the mainshock hypocenter, whose depths range from 15 to 25 km, shows a thrust mechanism, with strike = 300°, dip = 30.9° and rake = 77°. First arrival polarities from other aftershocks did not provide enough resolution to define

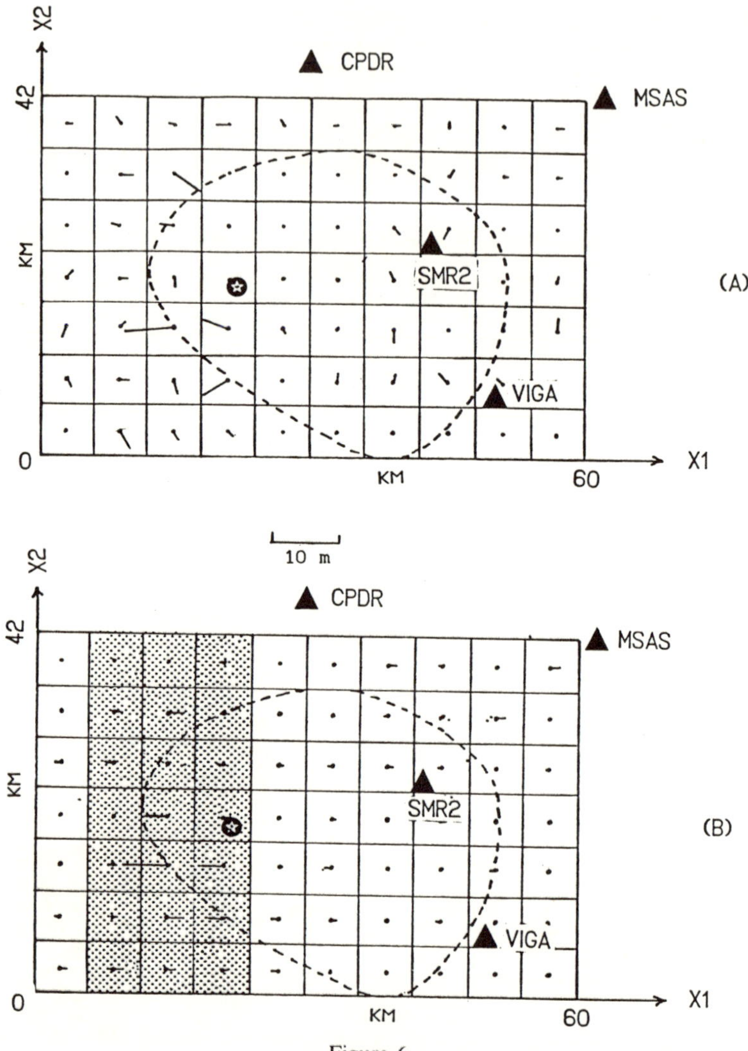

Figure 6

Slip distribution during the mainshock based on the inversion of near-field strong-motion records from GAA (see text). The mainshock location is shown by a star, solid triangles indicate stations used in the inversion. a) Unconstrained solution. b) Solution constraining motion in the $X1$ direction. Shaded area is the area assumed to contribute the most to observed ground motion. After SINGH *et al.* (1989).

a single or composite mechanism for other subvolumes. Even though the quality of the solution is poor, the result is similar to that reported for the mainshock, except for the dip which in the case of the mainshocks is shallower. It is also similar to the focal plane solution obtained by SINGH *et al.* (1990), through integrated near-field P-wave modeling, for the largest of the aftershocks which was located in the same subvolume.

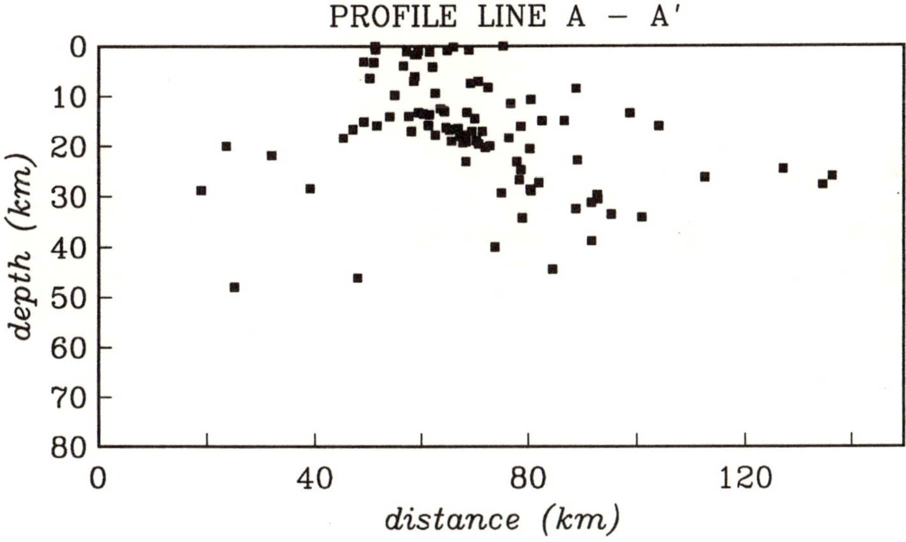

Figure 7

Distance vs. depth profile for aftershock hypocenters as projected on line *A-A′*. Distance is measured from trench.

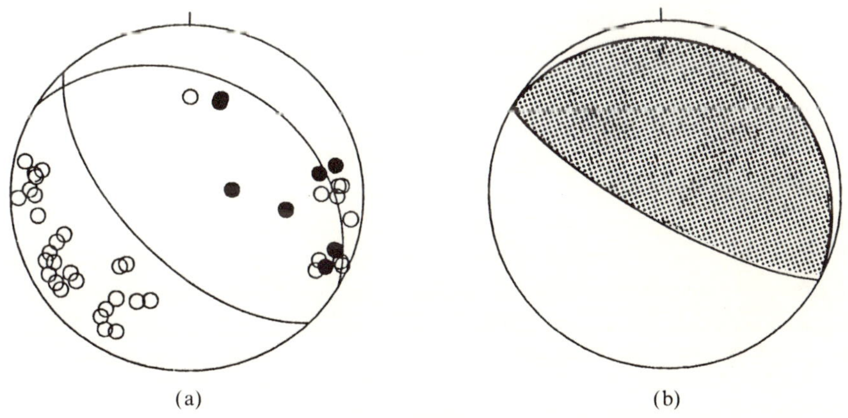

Figure 8

a) Composite focal mechanism solution for the San Marcos aftershocks. b) PDE focal mechanism solution for the April 25, 1989 mainshock.

Implications for the Future Occurrence of a Large Event in the Area

It is worth mentioning that the boundary between the ASM and OM segments, which could be taken as located in the area between the 1957 and 1982 rupture zones (Figure 4) and which in turn coincides with the eastern edge of the 1907 rupture, has been proposed as a major limit dividing the Mexican subduction zone into two different subduction geometries and physical characteristics (SINGH and

MORTERA, 1991). They based their assumption on several pieces of information including the number of source-time functions required to model teleseismic P-wave forms for events at either side, the ratio of surface- to body-wave seismic moment, as well as on other geophysical and morphological data. They propose that this limit could act as a barrier to rupture propagation and thus consider it as the SE limit of the actual Gerrero gap. This conclusion is also supported by the observation (ANDERSON *et al.*, 1989) that the 1982 Ometepec doublet exhibited greater affinity with Oaxaca events to the east and not with other Guerrero shocks, according to the relation between M_0 vs. M_S for large subduction events in Mexico.

Thus, we are faced with two scenarios: one, in which a major event ruptures the entire CG and ASM segments; the other is the possibility of both segments breaking in single independent events. As seen in Figure 4, one estimate of rupture length for the 1907 event (GONZÁLEZ-RUIZ and MCNALLY's, 1988) agrees with the length of the ASM segment. However, NISHENKO and SINGH's (1987) estimate (≈ 160 km) would involve partial rupture of the CG segment as well as all of the ASM. SINGH and MORTERA (1991) conclude that a magnitude $M_W = 8.2$ is reasonable if the entire CG and ASM regions were to break in a single event.

Following a similar analysis, we can estimate the maximum expected magnitude for the case of rupture of ASM alone. ASTIZ *et al.* (1987) have proposed the following relation between elapsed time since the last large event in a single region and average M_0, valid for Mexico and Central America subduction zones

$$\log T = \tfrac{1}{3} \cdot \log M_0 - 7.5$$

where T is in years and M_0 is in dyn-cm. If we follow ASTIZ *et al.*'s (1984) assumption, employed in the determination of the above relation, that the 1962 doublet formed part of the same strain release episode as the 1957 shock, and use the latter as the time of the last event, we can estimate a current moment deficit of about 1.0×10^{27} dyn-cm in the ASM region. This corresponds roughly to $M_S = 7.3$. The moment released by the 1989 event is within error margins of the latter estimate and thus cannot be used towards defining more precisely the expected deficit.

Conclusions

Aftershocks of the April 25, 1989 event delimit an area of about 780 km^2. One week aftershock activity shows a clustering of events near the mainshock epicenter which in turn coincides with the zone of maximum slip as determined from near-field accelerograms (SINGH *et al.*, 1989).

The aftershock area overlaps the aftershock zone of the 1957 earthquake and it may also coincide with the epicentral zone of the 1907 shock, based on its maximum intensity area.

A composite focal mechanism for aftershocks occurring near the mainshock hypocenter indicates a thrust fault (strike $= 300°$, dip $= 30.9°$ and rake $= 77°$) with similar characteristics to those of the mainshock.

The depth characteristics of the hypocenters define a low angle Benioff zone, with a slight tendency for the slab to follow a subhorizontal trajectory at a distance of about 110 km from the trench. This is similar to what is observed to the west in the neighboring Guerrero gap.

Even though in this region, as has occurred in the past, segments are likely to break independently from each other, the possibility of rupture continuing from the Guerrero gap to the neighboring San Marcos segment and thus the generation of another event breaking both segments (reaching a magnitude larger than 8.0) cannot be overlooked.

Acknowledgements

The authors wish to thank J. Estrada and T. Monfret for their assistance during field work. H. Mijares assisted with the data processing. T. Monfret, G. Suárez and S. K. Singh kindly revised the manuscript and offered useful suggestions. S. Jaume and an anonymous reviewer's comments greatly improved the content and presentation of the article. This work was supported by the Instituto de Geofísica and Instituto de Ingeniería (Project 9729), National University of Mexico.

References

ANDERSON, J. G., SINGH, S. K., ESPÍNDOLA, J. M., and YAMAMOTO, J. (1989), *Seismic Strain Release in the Mexican Subduction Thrust*, Phys. of the Earth and Plan. Int. *58*, 307–322.

ANDERSON, J. G., BODIN, P., BRUNE, J. N., PRINCE, J., SINGH, S. K., QUAAS, R., and ONATE, M. (1986), *Strong Ground Motion from the Michoacán, México, Earthquake*, Science *233*, 1043–1049.

ASTIZ, L., and KANAMORI, H. (1984), *An Earthquake Doublet in Ometepec, Guerrero, Mexico*, Phys. Earth and Plan. Int. *34*, 24–45.

ASTIZ, L., KANAMORI, H., and EISSLER, H. (1987), *Source Characteristics of Earthquakes in the Michoacán Seismic Gap in Mexico*, Bull. Seismol. Soc. Am. *77*, 1326–1346.

BÖSE, E., VILLAFANA, A., and GARCÍA, J. (1908), *El temblor del 14 de abril de 1907*, Parergones Inst. Geol. Mex. *2*, 135–258.

CHASE, C., MENARD, H., and MAMMERICK, J. (1970), *Bathymetry of the North Pacific*, Charts 1–6, Scripps Institute of Oceanography, Institute of Marine Resources, La Jolla, California.

EISSLER, H. K., and McNALLY, K. C. (1989), *Seismicity and Tectonics of the Rivera Plate and Implications for the 1932 Jalisco, Mexico, Earthquake*, J. Geophys. Res. *89*, 4520–4530.

GONZÁLEZ-RUIZ, J. R., and McNALLY, K. C. (1988), *Stress Accumulation and Release since 1882 in Ometepec, Guerrero, Mexico: Implications for Failure Mechanisms and Risk Assessments of a Seismic Gap*, J. Geophys. Res. *93*, 6297–6317.

GONZÁLEZ-RUIZ, J. R., and McNALLY, K. C. (1981), *Seismic Pattern before the Acapulco, Mexico, Earthquake ($M_S = 7.5$) of July 28, 1957*, EOS Trans., AGU *62*, 948.

HABERMANN, R. E. (1988), *Precursory Seismic Quiescence: Past, Present and Future*, Pure and Appl. Geophys. *126*, 279–318.

HARTZELL, S., and IIDA, M. (1990), *Source Complexity of the 1987 Whittier Narrows, California, Earthquake from the Inversion of Strong Motion Records*, J. Geophys. Res. *95*, 12475–12485.

LEE, W, and LAHR, J. C. (1978), *HYPO71 (Revised): A Computer Program for Determining Hypocenter, Magnitude and First Motion Pattern of Local Earthquakes*, U.S. Geological Survey, Open-File Report, 75-311.

MATUMOTO, T., OHTAKE, M., LATHAM, G., and UMANA, J. (1977), *Crustal Structure in Southern Central America*, Bull. Seismol. Soc. Am. *67*, 121–134.

MCCANN, W., NISHENKO, S. P., SYKES, L. R., and KRAUSE, J. (1979), *Seismic Gaps and Plate Tectonics: Seismic Potential for Major Boundaries*, Pure and Appl. Geophys. *117*, 1082–1147.

MCNALLY, K. C., and MINSTER, J. B. (1981), *Nonuniform Seismic Slip Rates along the Middle America Trench*, J. Geophys. Res. *86*, 4949–4959.

MINSTER, J. B., and JORDAN, T. H. (1978), *Present Day Plate Motions*, J. Geophys. Res. *83*, 5331–5354.

NISHENKO, S. P., and SINGH, S. K. (1987), *The Acapulco-Ometepec, Mexico, Earthquakes of 1907–1982: Evidence for a Variable Recurrence History*, Bull. Seismol. Soc. Am. *77*, 1359–1367.

PONCE, L., PARDO, M., COMTE, D., and MALAVÉ, G., *Spatio-temporal Analysis of the Guerrero Mexico Seismic Gap Seismicity (m > 4.5): A Quiescent Subregion Lasting 14 Years*, submitted to Bull. Seismol. Soc. Am.

PONCE, L., GAULON, R., SUÁREZ, G., and LOMAS, E. (1992), *Geometry and State of Stress of the Downgoing Cocos Plate in the Isthmus of Tehuantepec, Mexico*, Geophys. Res. Lett. *19*, 773–776.

SINGH, S. K., and MORTERA, F. (1991), *Source-time Functions of Large Mexican Subduction Earthquakes, Morphology of the Benioff Zone, and the Extent of the Guerrero Gap*, J. Geophys. Res. *96*, 21487–21502.

SINGH, S. K., ORDAZ, M., and NAVA, E. (1990), *Modeling of a Few Moderate Earthquakes Located at the Periphery of the Guerrero Gap Using Near-field Digital Seismograms*, EOS Trans., AGU *71*.

SINGH, S. K., ORDAZ, M., QUAAS, R., and MENA, E. (1989), *Estudio preliminar de la fuente del temblor del 25 de abril de 1989 ($M_S = 6.9$) a partir de los datos de movimientos fuertes*, Sociedad Mexicana de Ingeniería Sísmica, Memorias, Reunión Anual, Acapulco.

SINGH, S. K., ESPÍNDOLA, J. M., YAMAMOTO, J., and HAVSKOV, J. (1982), *Seismic Potential of the Acapulco–San Marcos Region Along the Mexican Subduction Zone*, Geophys. Res. Lett. *9*, 633–636.

SINGH, S. K., ASTIZ, L., and HAVSKOV, J. (1981), *Seismic Gaps and Recurrence Periods of Large Earthquakes Along the Mexican Subduction Zone: A Reexamination*, Bull. Seismol. Soc. Am. *71*, 827–843.

SUÁREZ, G., MONFRET, T., WITTLINGER, G., and DAVID, C. (1990), *Geometry of Subduction and Depth of the Seismogenic Zone in the Guerrero Gap, Mexico*, Nature *345*, 336–338.

SUÁREZ, G., LIGORRÍA, J. P., and PONCE, L. (1992), *Preliminary Crustal Structure of the Coast of Guerrero, Mexico Using the Minimum Apparent Velocity of Refracted Waves*, Geofís. Int. *31*, 247–252.

WALD, D. J., HELMBERGER, D. V., and HEATON, T. H. (1991), *Rupture Model of the 1989 Loma Prieta Earthquake from the Inversion of Strong-motion and Broadband Teleseismic Data*, Bull. Seismol. Soc. Am. *81*, 1540–1572.

(Received October 16, 1991, revised July 10, 1992, accepted September 10, 1992)

PAGEOPH, Vol. 140, No. 2 (1993)

0033–4553/93/020301–16$1.50 + 0.20/0
© 1993 Birkhäuser Verlag, Basel

Long-term Earthquake Prediction along the Western Coast of South and Central America Based on a Time Predictable Model

ELEFTHERIA E. PAPADIMITRIOU[1]

Abstract —The repeat times, T, of strong shallow mainshocks in fourteen seismogenic sources along the western coast of South and Central America have been determined and used in an attempt at long-term forecasting. The following relation was determined:

$$\log T = 0.22 M_{\min} + 0.21 M_p + a$$

between the repeat time, T, and the magnitudes, $M_{\min}$, of the minimum mainshock considered and M_p, of the preceding mainshock. No dependence of the magnitude, M_f, of the following mainshock on the preceding interevent time, T, was found. These results support the idea that the time-predictable model is valid for this region. This is an interesting property for earthquake prediction since it provides the ability to predict the time of occurrence of the next strong earthquake. A strong negative dependence of M_f on M_p was found, indicating that a large mainshock is followed by a smaller magnitude one, and *vice versa*.

The probability for the occurrence of the expected strong mainshocks ($M_s \geq 7.5$) in each of the fourteen seismogenic sources during the next 10 years (1992–2002) is estimated, adopting a lognormal distribution for earthquake interevent times. High probabilities ($P_{10} > 0.80$) have been calculated for the seismogenic sources of Oaxaca, Chiapas and Southern Peru.

Key words: Time-predictable model, probabilities, South and Central America region.

Introduction

Repeat times of large earthquakes have been extensively used during recent decades by various researchers as a tool for long-term earthquake prediction. Repeat time is taken to be the time interval between the largest shocks that occur at a given place along a plate boundary or in a specific seismic zone, and it varies both along and among seismic zones. In the time-predictable model, the time interval between large shocks depends on the size of the preceding event. This model is in accord with laboratory experiments on stick-slip behaviour on pre-existing faults (or surfaces) in that slip takes place in a given sample, when the stress reaches a level that is nearly constant among large events (SYKES, 1983). If the

---

[1] Laboratory of Geophysics, Aristotle University of Thessaloniki, Thessaloniki, Macedonia, GR54006, Greece.

stress drop or displacement at a given place vary from event to event, it is understandable that the time needed to restore stresses to that level by plate movements will vary from event to event. Thus, while the time intervals between large events may constitute a random distribution, the time-predictable model indicates that specific time intervals are related in a deterministic manner to the displacement in the preceding earthquakes.

SYKES and QUITTMEYER (1981) found that data on the geometry, seismic moment and repeat time of large shocks of both the strike-slip and convergent types agree better with the time-predictable model of earthquake recurrence than with the slip-predictable model. SHIMAZAKI and NAKATA (1980) found that the relative amounts of uplift in several large earthquakes off southwestern Japan are also in accord with the time-predictable model.

PAPAZACHOS (1988a,b, 1989, 1992), based on the repeat times of shallow mainshocks which occurred in the Aegean and the surrounding area, found that the time-predictable model and not the slip-predictable model holds for the seismogenic sources in which this area was divided. The author proved that this model can also be applied in seismogenic sources that may include several faults and where in addition to the maximum (characteristic) earthquake, other mainshocks are generated, and moreover that it holds for several seismotectonic environments.

It is the aim of the present study to apply the methodology suggested by the above-mentioned author, in the seismic zone along the western coast of South and Central America. This seismic zone was selected because it is homogeneous from the seismotectonic point of view (simple plate boundaries), and has been repeatedly struck by large events (adequate number of repeat times). In the examined area the subduction of the Nazca plate underneath the curving margin of western South America takes place, quite simple since it defines a single plane dipping about $30°$, and the subduction of the Cocos plate along the Middle America Trench, beneath the North American and Caribbean plates (MOLNAR and SYKES, 1969; DEAN and DRAKE, 1978; among others).

Method Applied

PAPAZACHOS (1988a,b, 1989, 1992), based on the repeat times of strong shallow earthquakes in the seismogenic sources in Greece, found that their logarithms are linearly correlated with the magnitudes, M_p, of the preceding mainshocks as well as with the magnitudes, M_f, of the following mainshocks. He found a strong positive correlation between $\log T$ and M_p, and a weak and negative correlation between $\log T$ and M_f. Therefore, the time-predictable model holds for the earthquakes of this region and not the slip-predictable model. A relation of the form

$$\log T_t = bM_{\min} + cM_p + a, \qquad (1)$$

has been determined, where M_{min} is the magnitude of the smallest mainshock in the data sample, "c" and "b" parameters having the same value for all the sources, while the parameter "a" has a different value in each source.

For the application of the above suggested methodology, the following conditions must be fulfilled. First, appropriate definition of seismogenic sources is needed, so as to have the same seismotectonic environment, and their individual characteristics as far as the behaviour of the seismic activity is concerned. Secondly, a reliable catalogue of earthquakes, that is, accurate, complete and homogeneous, must be used as a data source.

Following the procedure suggested by the above-mentioned author, the shallow ($h \leq 60$ km) earthquakes which occurred in each of the fourteen seismogenic sources, in which the whole seismic zone was divided, were depicted from a complete and homogeneous catalogue. From the data set corresponding in each seismogenic source, the after- and foreshocks were omitted in order to have a sample containing only mainshocks. We use the terms "aftershock" and "foreshock" in their broad sense because we are interested in applying a model which can predict the mainshocks in each seismogenic source, that is, the strong earthquakes which occur at the beginning and the end of each seismic cycle and not smaller earthquakes which occur during the preseismic and postseismic activations. In particular, we consider as foreshocks or aftershocks the earthquakes which precede or follow mainshocks with magnitudes 6.0–7.0, 7.1–8.0 and larger than 8.0 up to 9, 12.5 and 17 yrs, respectively. These time intervals are in accordance with PAPAZACHOS (1992) and the duration of the pre- and postseismic activity during the seismic cycle, found by KARAKAISIS et al. (1991).

The cumulative magnitude, M, of each sequence was estimated, that is, the magnitude which corresponds to the total moment released by the major shocks (mainshock and its foreshocks and aftershocks) of each sequence, according to the relation suggested by KANAMORI (1977). These cumulative magnitudes were used in this study instead of the magnitudes of the mainshocks.

We now have the mainshocks and the year of their occurrence for each seismogenic source. We need the interevent times between the mainshocks which have magnitudes larger or equal to a certain value, taking into consideration the completeness of the data for the certain time period. The determination of these interevent times, needed in the application of the present methodology, is performed in the following manner. The minimum magnitude mainshock of each sample is considered and the interevent times of the mainshocks which have magnitudes equal to or larger than this magnitude are estimated. This smallest magnitude is then ignored in the second step, the mainshocks which have magnitudes equal to or larger than the magnitude of the new smaller mainshock in the data sample are considered, and the interevent times of these shocks are now estimated. The same is repeated until the maximum magnitude mainshock in the data sample is determined, to obtain at least one interevent time.

Seismogenic Sources and Data Used

The division of South and Central America into five seismic zones suggested by LAY and KANAMORI (1981) and LAY *et al.* (1982) is first considered in the present study. This division was based on regional variations of fault lengths of the large earthquakes, on the degree of coupling and on the segmentation of the stress regime by transverse boundaries. In southern Chile, strong coupling on the fault plane overcomes the effect of lateral segmentation, and great earthquakes tend to recur regularly over approximately the same rupture zone, having length greater than 500 km. In Peru and Central Chile substantial aseismic slip and heterogeneous stress distribution generate moderate size, but complex ruptures, —weak coupling—, large earthquakes repeatedly rupture the same portion of the subduction zone, but without coalescing to generate large events, although the length scale of the subzones ranges from 100–300 km. In Colombia, the interaction between coupling and segmentation demonstrated temporal variation in the rupture mode, with occasionally very large ruptures (sometimes 500 km long), spanning segments of the trench that fail individually at other times. The Middle America seismic zone appears to have small-scale but fairly uniform stress distribution, producing an occasional clustering of events in time and space with smaller rupture dimensions (100–150 km).

Based on the above described general characteristics of the seismic behaviour in the five seismic zones, their division into seismogenic sources is made in the present study on the basis of spatial clustering of the epicenters of strong earthquakes, seismicity level, maximum earthquake observed, type of faulting and geomorphological criteria. The dimensions of the seismogenic sources are chosen according to the maximum "characteristic" rupture length observed in the corresponding source. The fit of the data of each source to the time-predictable model has been used as a supplementary criterion for this separation.

The Central America seismic zone is divided into six seismogenic sources (having lengths 100–150 km) based on the spatial distribution of the aftershock areas of large ($M_s \geq 7.0$) shallow earthquakes (ASTIZ *et al.*, 1987; among others).

The seismogenic sources of North Colombia, South Colombia, Northern Peru, Southern Peru, North Chile, Central Chile and Valparaiso are about 500 km long, in accordance with the maximum rupture lengths reported for the 31 January 1906 earthquake (500 km) or the 1922 Atacama Desert earthquake (more than 400 km). Rupture lengths of this order have also been reported for earthquakes occurring before the present century, but with less reliability.

Southern Chile is considered as one seismogenic source, although its length is long enough. The criterion on which this definition is based is that the rupture length of the great 1960 Chilean earthquake was almost equal to 1000 km.

Figures 1a and 1b show the definition of the fourteen seismogenic sources, along with the epicentral distribution of earthquakes which have magnitudes larger than or equal to 7.0. The mainshocks are denoted by black circles, while the fore- and aftershocks are signified by open ones.

Information regarding the surface wave magnitudes and the epicenters of the earthquakes used was taken from a worldwide catalogue compiled by TSAPANOS (1985) and TSAPANOS *et al.* (1990), which covers the time period 1898–1985 and includes events with $M_s \geq 5.5$. This catalogue is complete for the following time periods and their corresponding magnitude ranges: 1898–1985, $M_s \geq 7.0$; 1931–1985, $M_s \geq 6.5$; 1951–1985, $M_s \geq 5.0$ and 1966–1985, $M_s \geq 5.5$. From this catalogue only the events with $M_s \geq 6.0$ have been considered in the present study. This catalogue was checked and corrected according to the catalogue of PACHECO and SYKES (1992).

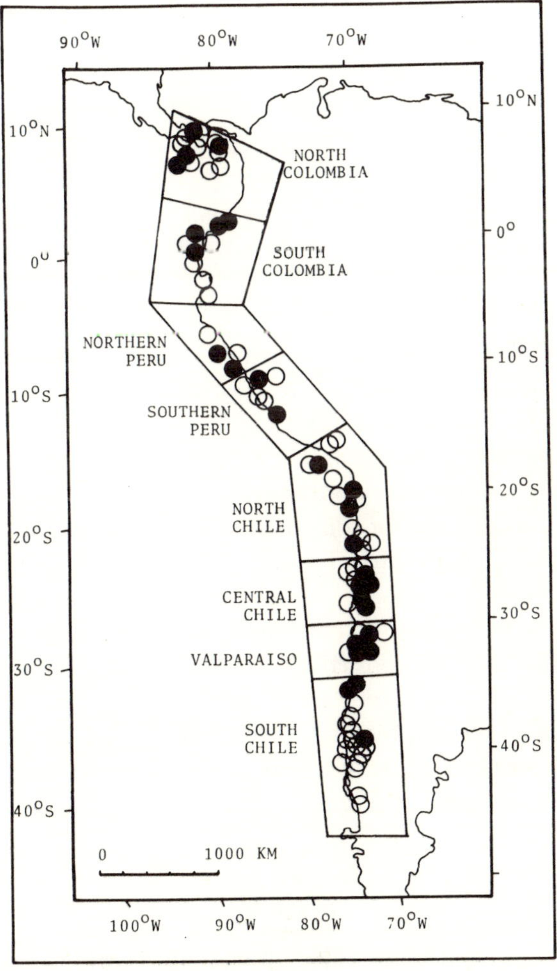

Figure 1(a)

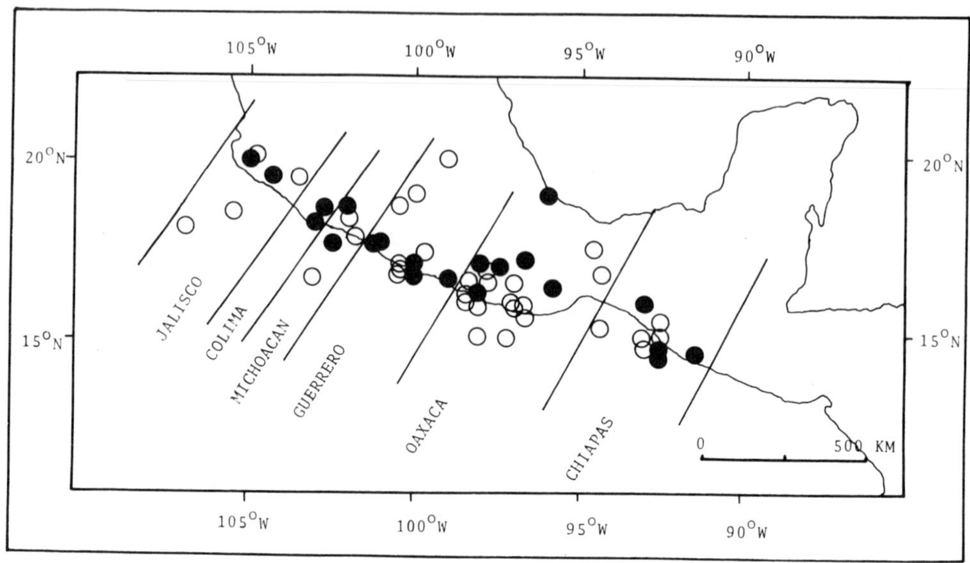

Figure 1(b)

Figure 1

The seismogenic sources in which the seismic zones of South America (1a) and Central America (1b) were divided. The epicentral distribution of earthquakes with $M_s \geq 7.0$ is also shown. Black circles represent the mainshocks and open circles the fore- and aftershocks.

Information on the mainshocks for which interevent times have been determined is given in Table 1. In addition to this, all the events with $M_s \geq 7.0$ which are considered as foreshocks (indicated by f) or aftershocks (indicated by a), are also shown. In the first column of this table a name is written for each seismogenic source. The second, third and fourth columns give the date, epicenter and the surface wave magnitude for the mainshocks that satisfy the completeness condition. The fifth column provides the cumulative magnitudes, M, of each seismic sequence.

Table 2 supplies the values of the minimum magnitude, $M_{\min}$, of the mainshock, the magnitude of the preceding mainshock, M_p, the magnitude of the following mainshock, M_f, the interevent (repeat) time, T, between the two shocks, the year of occurrence of the preceding mainshock, t_p, and the year of occurrence of the following mainshock, t_f, as these values were derived from the basic data of Table 1 for each seismogenic source.

Results

The procedure suggested by PAPAZACHOS (1992) was followed here in order to estimate the parameters of the relation (1) between the repeat time, T, and the

Table 1

Information on the basic data used for each seismogenic source

Seismogenic Source	Date	Epicenter		M_s	M
SOURCE 1	20 1 1900	20.0	-105.0	7.3	f
JALISCO	16 12 1900	20.0	-105.0	7.4	7.6
	16 11 1925	18.0	-107.0	7.0	f
	3 6 1932	19.5	-104.2	8.0	8.1
	18 6 1932	19.5	-103.5	7.6	a
	30 11 1934	18.5	-105.5	7.0	a
	21 10 1950	17.8	-105.5	6.7	6.7
	27 3 1960	19.0	-105.0	6.1	6.3
	18 10 1973	19.4	-105.0	6.3	6.3
SOURCE 2	15 4 1941	18.0	-103.0	7.5	7.5
COLIMA	30 1 1973	18.5	-102.9	7.3	7.3
SOURCE 3	7 6 1911	17.5	-102.5	7.6	7.6
MICHOACAN	20 1 1955	15.5	-104.4	6.2	6.2
	25 10 1981	18.2	-102.0	7.3	f
	19 9 1985	18.5	-102.3	8.1	8.3
	21 9 1985	17.8	-101.7	7.6	a
	30 4 1986	18.4	-103.0	7.0	a
SOURCE 4	15 4 1907	17.0	-100.0	7.6	7.9
GUERRERO	30 7 1909	17.0	-100.5	7.2	a
	16 12 1911	17.0	-100.5	7.4	a
	15 1 1913	19.0	-100.0	7.5	a
	14 6 1913	20.0	-99.0	7.5	a
	17 8 1929	16.3	-99.0	7.0	7.0
	22 2 1943	17.7	-101.5	7.3	7.4
	3 10 1947	18.8	-100.7	7.0	a
	17 11 1950	16.8	-100.7	7.0	f
	27 7 1957	16.7	-99.0	7.5	7.8
	11 5 1962	17.3	-99.6	7.0	a
	14 3 1979	17.8	-101.3	7.4	7.5
SOURCE 5	24 1 1899	17.0	-98.0	7.9	8.0
OAXACA	14 1 1903	15.0	-98.0	7.6	a
	30 3 1914	19.0	-96.0	7.5	7.7
	29 12 1917	15.0	-97.0	7.5	a
	22 3 1928	16.0	-96.0	7.3	f
	17 4 1928	17.5	-94.5	7.0	f
	17 6 1928	16.2	-98.0	7.6	8.1
	4 8 1928	16.0	-97.0	7.2	a
	9 10 1928	16.0	-97.0	7.4	a
	15 1 1931	16.0	-96.7	7.6	a
	23 12 1937	16.7	-98.5	7.3	a
	15 5 1946	15.5	-96.7	7.0	f
	6 1 1948	17.0	-98.0	7.0	f
	21 9 1949	16.9	-94.2	7.0	f
	14 12 1950	17.0	-97.5	7.1	7.7
	12 12 1951	16.5	-96.9	7.0	a
	23 8 1965	16.3	-95.8	7.6	7.8
	2 8 1968	16.6	-97.8	7.2	a

Table 1 (*Contd*)

Seismogenic Source	Date	Epicenter		M_s	M
	29 11 1978	16.1	−96.6	7.6	7.8
	7 6 1982	16.5	−98.3	7.0	a
	7 6 1982	16.6	−98.3	7.0	a
SOURCE 6 CHIAPAS	23 9 1902	16.0	−93.0	7.7	7.7
	10 12 1925	15.5	−92.5	7.0	f
	14 12 1935	14.7	−92.5	7.2	7.7
	11 2 1941	15.2	−94.4	7.0	a
	15 6 1943	14.6	−93.0	7.0	a
	28 6 1944	15.0	−92.5	7.1	f
	22 12 1949	15.9	−93.0	7.0	f
	23 10 1950	14.5	−91.5	7.2	7.6
	29 4 1970	14.7	−92.6	7.1	7.1
SOURCE 7 NORTH COLOMBIA	20 1 1904	7.0	−79.0	7.1	f
	20 12 1904	8.5	−83.0	7.1	f
	2 10 1913	7.5	−79.0	7.5	7.8
	20 4 1914	7.6	−79.1	7.5	a
	10 8 1927	8.0	−81.5	7.0	f
	18 7 1934	8.0	−82.5	7.4	7.8
	5 12 1941	8.5	−83.0	7.3	a
	2 5 1943	6.5	−80.0	7.0	a
	16 5 1952	6.7	−78.9	7.0	f
	26 7 1962	7.5	−82.8	7.1	7.8
	18 9 1962	7.5	−82.3	7.0	a
	3 4 1983	8.8	−83.1	7.3	7.5
SOURCE 8 SOUTH COLUMBIA	7 1 1901	−2.0	−82.0	7.1	f
	31 1 1906	1.0	−81.5	8.1	8.1
	30 6 1916	0.5	−82.0	7.0	a
	14 5 1942	−0.7	−81.5	7.7	8.0
	14 5 1942	0.0	−80.0	7.2	a
	12 12 1953	−3.4	−80.6	7.4	f
	19 1 1958	1.5	−79.5	7.3	7.8
	10 12 1970	−4.0	−80.7	7.4	f
	12 12 1979	1.6	−79.4	7.6	7.9
SOURCE 9 NORTHERN PERU	21 6 1937	−8.5	−80.9	7.1	7.4
	10 11 1946	−8.5	−77.5	7.1	a
	10 11 1960	−6.9	−80.8	7.0	f
	31 5 1970	−9.2	−78.8	7.6	7.7
	21 8 1985	−9.2	−78.9	6.0	6.0
SOURCE 10 SOUTHERN PERU	24 5 1940	−10.5	−77.0	7.7	f
	24 8 1942	−15.0	−76.0	8.0	8.2
	1 11 1947	−10.5	−75.0	7.1	a
	17 10 1966	−10.7	−78.6	7.8	8.1
	3 10 1974	−12.2	−77.6	7.6	a
	9 11 1974	−12.4	−77.5	7.0	a
	19 11 1982	−10.6	−74.7	6.5	6.5
SOURCE 11 NORTH CHILE	30 8 1906	−21.0	−70.0	7.0	f
	6 9 1910	−25.0	−70.0	7.0	f
	15 9 1911	−20.0	−72.0	7.0	f
	28 7 1913	−17.0	−74.0	7.0	f

Table 1 (*Contd*)

Seismogenic Source	Date	Epicenter		M_s	M
	6 8 1913	−17.0	−74.0	7.7	7.8
	6 1 1922	−16.5	−73.0	7.0	a
	11 10 1922	−16.0	−72.5	7.1	a
	23 2 1933	−20.0	−71.0	7.1	7.4
	13 7 1936	−24.5	−70.0	7.0	a
	29 7 1957	−23.3	−71.1	7.0	f
	2 12 1960	−24.4	−70.0	7.1	f
	21 12 1967	−21.9	−70.1	7.3	7.8
	5 3 1987	−24.4	−70.2	7.3	7.5
	12 4 1988	−17.2	−72.3	7.0	a
SOURCE 12 CENTRAL CHILE	7 12 1903	−27.0	−70.0	7.3	7.5
	8 6 1909	−26.5	−70.5	7.2	a
	4 12 1918	−26.0	−71.0	7.4	f
	7 11 1922	−28.0	−72.0	7.1	f
	11 11 1922	−28.5	−70.0	8.1	8.2
	15 5 1925	−26.0	−71.5	7.1	a
	2 8 1946	−26.5	−70.5	7.1	7.3
	18 12 1956	−25.4	−70.1	7.0	f
	28 12 1966	−25.5	−70.7	7.7	7.8
	4 10 1983	−26.6	−70.6	7.3	7.3
SOURCE 13 VALPARAISO	17 8 1906	−33.0	−72.0	8.0	8.0
	18 3 1906	−32.5	−72.0	7.1	f
	6 4 1943	−30.7	−72.0	7.7	7.8
	15 1 1944	31.2	68.7	7.0	a
	4 9 1958	−33.5	−69.5	6.8	6.8
	9 7 1971	−32.5	−71.2	7.7	7.7
	16 10 1981	−33.2	−73.1	7.2	f
	3 3 1985	−33.1	−71.7	7.8	8.0
	9 4 1985	−34.1	−71.6	7.2	a
SOURCE 14 SOUTH CHILE	30 1 1914	−35.0	−73.0	7.4	7.5
	2 3 1919	−41.0	−73.5	7.1	a
	2 3 1919	−41.0	−73.5	7.0	a
	10 12 1920	−39.0	−73.0	7.1	f
	21 11 1927	−44.5	−73.0	7.0	f
	1 12 1928	−35.0	−72.0	7.8	8.1
	25 1 1939	−36.2	−72.2	7.6	a
	11 10 1940	−41.5	−74.5	7.0	a
	6 5 1953	−36.5	−72.6	7.4	f
	21 5 1960	−37.6	−73.1	7.9	f
	22 5 1960	−37.7	−72.7	7.2	f
	22 5 1960	−38.1	−72.7	7.8	f
	22 5 1960	−38.1	−72.2	8.5	9.0
	22 5 1960	−38.2	−72.6	8.5	a
	6 6 1960	−45.7	−73.1	7.1	a
	20 6 1960	−38.2	−72.8	7.1	a
	20 6 1960	−39.1	−73.1	7.0	a
	14 2 1962	−37.8	−72.5	7.3	a
	1 6 1982	−41.4	−75.0	6.2	6.3

Table 2

Data used to determine the parameters of the empirical relations

Seismogenic Source	M_{min}	M_p	M_f	T	t_p	t_f
SOURCE 1	6.3	6.3	6.3	13.56	1960	1973
JALISCO	6.7	8.1	6.7	18.38	1932	1950
	7.6	7.6	8.1	31.46	1900	1932
SOURCE 2						
COLIMA	7.3	7.5	7.3	31.79	1941	1973
SOURCE 3	6.2	6.2	8.3	30.67	1955	1985
MICHOACAN	7.6	7.6	8.3	74.28	1911	1985
SOURCE 4	7.0	7.9	7.0	22.34	1907	1929
GUERRERO		7.0	7.4	13.51	1929	1943
		7.4	7.8	14.43	1943	1957
		7.8	7.5	21.63	1957	1979
	7.4	7.9	7.4	35.85	1907	1943
		7.4	7.8	14.43	1943	1957
		7.8	7.5	21.63	1957	1979
	7.5	7.9	7.8	50.29	1907	1957
		7.8	7.5	21.63	1957	1979
	7.8	7.9	7.8	50.29	1907	1957
SOURCE 5	7.7	8.0	7.7	15.18	1899	1914
OAXACA		7.7	8.1	14.21	1914	1928
		8.1	7.7	22.49	1928	1950
		7.7	7.8	14.69	1950	1965
		7.8	7.8	13.27	1965	1978
	7.8	8.0	8.1	29.40	1899	1928
		8.1	7.8	37.18	1928	1965
		7.8	7.8	13.27	1965	1978
	8.0	8.0	8.1	29.40	1899	1928
SOURCE 6	7.1	7.7	7.7	33.23	1902	1935
CHIAPAS		7.7	7.6	14.86	1935	1950
		7.6	7.1	19.52	1950	1970
	7.6	7.7	7.7	33.23	1902	1935
		7.7	7.6	14.86	1935	1950
	7.7	7.7	7.7	33.23	1902	1935
SOURCE 7	7.5	7.8	7.8	20.79	1913	1934
NORTH		7.8	7.8	28.02	1934	1962
COLOMBIA		7.8	7.5	20.69	1962	1983
	7.8	7.8	7.8	20.79	1913	1934
		7.8	7.8	28.02	1934	1962
SOURCE 8	7.8	8.1	8.0	36.29	1906	1942
SOUTH		8.0	7.8	16.68	1942	1958
COLOMBIA		7.8	7.9	21.90	1958	1979
	7.9	8.1	8.0	36.29	1906	1942
		8.0	7.9	37.58	1942	1979
	8.0	8.1	8.0	36.29	1906	1942

Table 2 (*Contd*)

Seismogenic Source	M_{min}	M_p	M_f	T	t_p	t_f
SOURCE 9	6.0	7.7	6.0	15.22	1970	1985
NORTHERN	7.4	7.4	7.7	32.94	1937	1970
PERU						
SOURCE 10	6.5	8.2	8.1	24.15	1942	1966
SOUTHERN		8.1	6.5	16.09	1966	1982
PERU	8.1	8.2	8.1	24.15	1942	1966
SOURCE 11	7.4	7.8	7.4	19.54	1913	1933
NORTH		7.4	7.8	34.83	1933	1967
CHILE		7.8	7.5	19.21	1967	1987
	7.5	7.8	7.8	54.37	1913	1967
		7.8	7.5	19.21	1967	1987
	7.8	7.8	7.8	54.37	1913	1967
SOURCE 12	7.3	7.5	8.2	18.93	1903	1922
CENTRAL		8.2	7.3	23.73	1922	1946
CHILE		7.3	7.8	20.40	1946	1966
		7.8	7.3	16.77	1966	1983
	7.5	7.5	8.2	18.93	1903	1922
		8.2	7.8	44.13	1922	1966
	7.8	8.2	7.8	44.13	1922	1966
SOURCE 13	6.8	7.8	6.8	15.41	1943	1958
VALPARAISO		6.8	7.7	12.84	1958	1971
		7.7	8.0	13.65	1971	1985
	7.7	8.0	7.8	36.64	1906	1943
		7.8	7.7	28.26	1943	1971
		7.7	8.0	13.65	1971	1985
	7.8	8.0	7.8	36.64	1906	1943
		7.8	8.0	41.91	1943	1985
	8.0	8.0	8.0	78.55	1906	1985
SOURCE 14	6.3	9.0	6.3	22.02	1960	1982
SOUTH	7.5	7.5	8.1	14.84	1914	1928
CHILE		8.1	9.0	31.48	1928	1960
	8.1	8.1	9.0	31.48	1928	1960

magnitudes M_{min} and M_p. For this purpose the data listed in Table 2 were used and the least squares' differential correction was applied. These data concern all 14 seismogenic sources, with a total of 73 repeat times. The values

$$b = 0.22, \quad c = 0.21,$$

were found for all the data set and for the parameter a, one different value for each seismogenic source. The corresponding values of the standard deviation and the multiple correlation coefficient are 0.13 and 0.76, respectively. The positive correlation between the repeat time and the magnitude of the preceding mainshock indicates that the time-predictable model holds. A relation of the form

$$\log T = b'M_{\min} + c'M_f + a', \tag{2}$$

was tried between the repeat time, T, the magnitude, M_f, of the following mainshock and the magnitude, $M_{\min}$, of the smallest mainshock considered. The values $b' = 0.29$ and $c' = -0.02$ were found, with a standard deviation equal to 0.15 and a multilinear correlation coefficient equal to 0.65. The very low value of the parameter c' indicates no dependence of the magnitude of the following mainshock on the time elapsed since the previous mainshock, that is, the slip-predictable model does not hold.

A relation of the form

$$M_f = BM_{\min} + CM_p + m, \tag{3}$$

has also been tried, with the aim to find the relation between the magnitude, M_f, of the following mainshock, and the magnitudes of the preceding, M_p, and the minimum, $M_{\min}$, mainshocks. The values $B = 0.85$ and $C = -0.45$ were obtained. A standard deviation equal to 0.29 and a multilinear correlation coefficient equal to 0.79 were determined. The negative value of C means that large mainshocks are followed by small ones and *vice versa*. PAPAZACHOS (1992) also arrived at the same conclusion for earthquakes in Greece.

Prediction of Strong Shallow Earthquakes

Time-dependent models, which take into account both the recurrence time and the amount of time elapsed since a previous shock, are compatible with the seismic gaps hypothesis which suggests that the probability for a future earthquake to fill up the gap is small immediately following the occurrence of a characteristic earthquake and increases with the time elapsed since the previous event. These models are distinct from time-independent models (i.e., Poisson distributions) in that the later ones use only information concerning recurrence intervals and do not take into account the amount of time elapsed since a prior event.

The time-predictable model is in accordance with the previous mentioned hypothesis, suggesting that the larger the magnitude of the previous event, the longer will be the time until the next. This is expressed by the relation (1), from which we can predict the time of occurrence of the next mainshock which is larger than a certain value in each of the seismogenic sources, since all three parameters are known and the magnitude and time of the preceding mainshock are also known. There is, however, a scattering between the theoretical and observed repeat times. For this reason it may be better to determine the probability, P, of occurrence of an earthquake larger than a certain magnitude (e.g., ≥ 7.5) in a future certain time

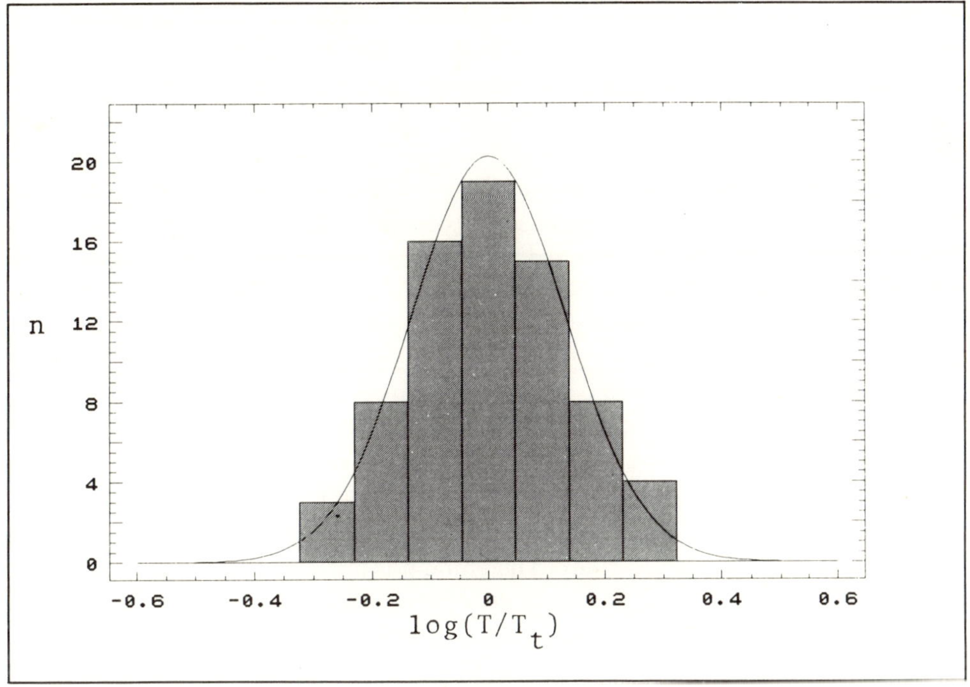

Figure 2
The frequency distribution of the observed repeat time compared to the theoretical one.

interval. The use of probability density functions or distributions, $f(t)$, of the data used must be known in order to proceed in such determinations. Once the distribution function is defined, earthquake recurrence time estimates can be presented in terms of a conditional probability, which describes the likelihood of failure within a given time interval, $t + dt$, provided the event has not occurred prior to time t.

NISHENKO and BULAND (1987) found that the lognormal distribution of T/T_m, where T are the observed repeat times for characteristic earthquakes of a certain seismic source and T_m their average, provides better fit than the more commonly used Gaussian and Weibull distributions. PAPAZACHOS (1988b, 1991) has also found that the lognormal distribution of T/T_t fits better the data coming from all the seismogenic sources of shallow earthquakes in Greece.

Figure 2 shows the frequency function for the quantity $\log(T/T_t)$, concerning the data for all the seismogenic sources examined here and the theoretical normal distribution with a mean equal to zero and a standard deviation equal to 0.13. The application of chi-square and Kolmogorov–Smirnoff tests support this distribution for our data sample. The normal and the Weibull distributions were also tested for the quantity T/T_t, and found less appropriate.

Previous research concerning the estimation of the potential of certain parts of plate boundaries around the circum-Pacific region to rupture in future large earthquakes has been undertaken by various researchers (McCann *et al.*, 1979; Nishenko, 1985; Nishenko, 1991). McCann *et al.* (1979), estimated this potential on the basis of the seismic gaps concept. They classified segments of plate boundaries into six categories of seismic potential for large earthquakes for the next few decades, and successfully forecasted the occurrence of large earthquakes in this region. Nishenko (1985) estimated the seismic potential for large and great earthquakes along the Chilean and southern Peruvian margins, based on a statistical analysis of historic repeat time data employing Weibull distribution and deterministic estimates of recurrence times based on the time-predictable model of earthquake recurrence. The same author (Nishenko, 1991) presented the seismic potential for 96 segments of simple plate boundaries around the circum-Pacific region, in terms of conditional probability for the occurrence of either large or great interplate earthquakes during the next 5, 10, and 20 years. The segments that Nishenko considered are quite different in the dimensions than the seismogenic sources defined in the present study. Therefore, it is impossible to have a one to one comparison of his results with the results of the present study.

Taking into account that the lognormal distribution (with mean equal to zero, and standard deviation equal to 0.13) holds for each seismogenic source and considering the time of occurrence of the preceding mainshock, the probabilities, P_{10}, of occurrene of the next large mainshocks ($M_s \geq 7.5$) during the next 10 years (1992–2002) were calculated and presented in Table 3.

Table 3

Probability of occurrence, P_{10}, of a mainshock with $M_s \geq 7.5$, in each seismogenic source during the next ten years (1992–2002). a is the constant of relation (1)

Sesimogenic Source	a	P_{10}
SOURCE 1—JALISCO	−1.81	0.70
SOURCE 2—COLIMA	−1.74	0.50
SOURCE 3—MICHOACAN	−1.35	0.00
SOURCE 4—GUERRERO	−1.91	0.61
SOURCE 5—OAXACA	−2.15	0.93
SOURCE 6—CHIAPAS	−1.93	0.87
SOURCE 7—NORTH COLOMBIA	−2.01	0.68
SOURCE 8—SOUTH COLOMBIA	−2.02	0.68
SOURCE 9—NORTHERN PERU	−1.77	0.46
SOURCE 10—SOUTHERN PERU	−2.00	0.81
SOURCE 11—NORTH CHILE	−1.86	0.08
SOURCE 12—CENTRAL CHILE	−1.95	0.63
SOURCE 13—VALPARAISO	−1.92	0.09
SOURCE 14—SOUTH CHILE	−2.02	0.60

Discussion

The time-predictable model is valid for several seismically active regions of the world, as it has been found by various researchers. The results obtained in the present study also favour the validity of the time-predictable model in the seismogenic sources of the southeastern Pacific. Certainly, it is ascertained that the dependence of the repeat time, T, on the magnitude of the preceding mainshock, M_p, is strong ($c = 0.21$), while the dependence of T on the magnitude of the following mainshock, M_f, is very weak ($c = -0.02$). This reveals that the slip-predictable model does not hold for the area under study. The strong negative dependence of the magnitude of the following mainshock, M_f, on the magnitude of the preceding mainshock, M_p ($C = -0.45$), shows that a small mainshock in a certain seismic zone is more likely to be followed by a large mainshock and *vice versa*.

Taking into consideration that the time-predictable model holds, the observed, T, and theoretical, T_t, repeat times were used in an attempt at long-term earthquake prediction. The lognormal distribution of the ratio T/T_t, appropriate for our data sample, and the time elapsed since the last mainshock were taken into account in order to estimate the probability of occurrence of the next mainshock. During the next ten years the probability estimated for the generation of an earthquake with $M_s \geq 7.5$ is very high in Oaxaca (0.93). High probabilities were calculated for Chiapas and Southern Peru (0.87 and 0.81, respectively). Intermediate values ($0.46 \leq P_{10} \leq 0.70$) have been obtained for the seismogenic sources of Jalisco, Colima, Guerrero, North Colombia, South Colombia, Northern Peru, Central Chile and South Chile. The probabilities estimated for the seismogenic sources of Jalisco, South Colombia and Central Chile are in accordance with the classification by MCCANN *et al.* (1979), who have characterized these areas as seismic gaps of category 2. Finally, very low values have been found ($0.00 \leq P_{10} \leq 0.09$) for Michoacán, North Chile and Valparaiso. The results concerning the seismogenic sources of Central America are in accordance with NISHENKO and SINGH (1987), who estimated time-dependent conditional probabilities for the recurrence of large and great earthquakes in this area.

Acknowledgements

The author would like to express her sincere appreciation to Prof. B. C. Papazachos for his creative ideas, his suggestions and the critical reading of the manuscript. Gratitude is also extended to Drs. E. Scordilis, C. Papazachos and C. Papaioannou for kindly providing the computer programs.

REFERENCES

ASTIZ, L., KANAMORI, H., and EISSLER, H. (1987), *Source Characteristics of Earthquakes in the Michoacán Seismic Gap in Mexico*, Bull. Seismol. Soc. Am. *77*, 1326–1346.

DEAN, B. W., and DRAKE, C. L. (1978), *Focal Mechanism Solutions and Tectonics of the Middle America Arc*, J. Geol. *86*, 11–128.

KANAMORI, H. (1977), *The Energy Release in Great Earthquakes*, J. Geophys. Res. *82*, 2981–2987.

KARAKAISIS, G. F., KOUROUZIDIS, M. C., and PAPAZACHOS, B. C. (1991), *Behaviour of seismic activity during a single seismic cycle*, In *Earthquake Prediction: State-of-the-Art*, Strasbourg, France, 15–18 October 1991, pp. 47–54.

KELLEHER, J., SYKES, L. R., and OLIVER, J. (1973), *Possible Criteria for Predicting Earthquake Locations and their Application to Major Plate Boundaries of the Pacific and the Caribbean*, J. Geophys. Res. *78*, 2547–2585.

LAY, T., and KANAMORI, H., *An asperity model of large earthquakes*. In *Earthquake Prediction*, An International Review (Simpson, D. W., and Richards, P. G. eds.) (Maurice Ewing Series, Amer. Geophys. Union 1981) vol. 4, pp. 579–592.

LAY, T., KANAMORI, H., and RUFF, L. (1982), *The Asperity Model and the Nature of Large Subduction Zone Earthquakes*, Earthq. Pred. Res. *1*, 3–71.

MCCANN, W. R., NISHENKO, S. P., SYKES, L. R., and KRAUSE, J. (1979), *Seismic Gaps and Plate Tectonics: Seismic Potential for Major Plate Boundaries*, Pure and Appl. Geophys. *117*, 1082–1147.

MOLNAR, P., and SYKES, L. R. (1969), *Tectonics of the Caribbean and Middle American Regions from Local Mechanisms and Seismicity*, Bull. Geol. Soc. Am. *80*, 1639–1684.

NISHENKO, S. P. (1985), *Seismic Potential for Large and Great Interplate Earthquakes along the Chilean and Southern Peruvian Margins of South America: A Quantitative Reappraisal*, J. Geophys. Res. *90*, 3589–3615.

NISHENKO, S. P. (1991), *Circum-Pacific Seismic Potential: 1989–1999*, Pure and Appl. Geophys. *135*, 169–259.

NISHENKO, S. P., and BULAND, R. (1987), *A Generic Recurrence Interval Distribution for Earthquake Forecasting*, Bull. Seismol. Soc. Am. *77*, 1382–1399.

NISHENKO, S. P., and SINGH, S. K. (1987), *Conditional Probabilities for the Recurrence of Large and Great Interplate Earthquakes along the Mexican Subduction Zone*, Bull. Seismol. Soc. Am. *77*, 2095–2114.

PACHECO, J. F., and SYKES, L. R. (1992), *Seismic Moment Catalog of Large Shallow Earthquakes, 1900 to 1989*, Bull. Seismol. Soc. Am. *82*, 1306–1349.

PAPAZACHOS, B. C. (1988a), *Seismic Hazard and Long-term Earthquake Prediction in Greece*, European School of Earthquake Sciences, Course on Earthquake Hazard Assessment, Athens, 9–16 May 1988, pp. 1–10.

PAPAZACHOS, B. C. (1988b), *Long-term Prediction of Earthquakes in Seismogenic Sources of Greece*, United Nations Seminar on the Prediction of Earthquakes, Lisbon, Portugal, 14–18 November 1988, pp. 1–10.

PAPAZACHOS, B. C. (1989), *A Time-predictable Model for Earthquake Generation in Greece*, Bull. Seismol. Soc. Am. *79*, 77–84.

PAPAZACHOS, B. C. (1991), *Long-term Earthquake Prediction in South Balkan Region Based on a Time-dependent Seismicity Model*, 1st General Conference of the B.P.U., Thessaloniki, September 26–28, 1991, pp. 1–7.

PAPAZACHOS, B. C. (1992), *A Time- and Magnitude-predictable Model for Generation of Shallow Earthquakes in the Aegean Area*, Pure and Appl. Geophys. *138*, 287–308.

SYKES, L. R. (1983), *Predicting great earthquakes*. In *Earthquakes: Observation, Theory and Interpretation* (Kanamori, H., and Boschi, E. eds.) pp. 398–435.

SYKES, L. R., and QUITTMEYER, R. C., *Repeat times of great earthquakes along simple plate boundaries*. In *Earth Prediction*, An International Review (Simpson, D. W., and Richards, P. G. eds.) (Maurice Ewing Series, Amer. Geophys. Union 1981) vol. 4, pp. 217–247.

SHIMAZAKI, K., and NAKATA, T. (1980), *Time-predictable Recurrence Model for Large Earthquakes*, Geophys. Res. Lett. *7*, 279–282.

TSAPANOS, T. M. (1985), *A Contribution to the Study of the Seismicity of the Earth*, Ph.D. Thesis, Univ. of Thessaloniki, pp. 1–147 (in Greek).

TSAPANOS, T. M., SCORDILIS, E. M., and PAPAZACHOS, B. C. (1990), *Global Seismicity during the Time Period 1966–1985*, Proc. XXII Gen Ass. Europ. Seism. Comm., Barcelona, 17–22 September 1990, pp. 709–714.

(Received November 19, 1992, revised February 25, 1993, accepted March 6, 1993)

PAGEOPH, Vol. 140, No. 2 (1993)

0033–4553/93/020317–14$1.50 + 0.20/0

Spatio-temporal Variations of Seismicity in the Southern Peru and Northern Chile Seismic Gaps

DIANA COMTE[1,2] and GERARDO SUÁREZ[1]

Abstract—The spatio-temporal variation of seismicity in the southern Peru and northern Chile seismic gaps is analyzed with teleseismic data ($m_b \geq 5.5$) between 1965 and 1991, to investigate whether these gaps present the precursory combination of compressional outer-rise and tensional downdip events observed in other subduction zones. In the outer-rise and the inner-trench (0 to 100 km distance from the trench) region, lower magnitude ($5.0 \leq m_b < 5.5$) events were also studied. The results obtained show that the gaps in southern Peru and northern Chile do not present compressional outer-rise events. However, both gaps show a continuous, tensional downdip seismicity. For both regions, the change from compressional to tensional regime along the slab occurs at a distance of about 160 km from the trench, apparently associated with the coupled-uncoupled transition of the interplate contact zone. In southern Peru, an increase of compressional seismicity near the interplate zone and of tensional events ($5.0 \leq m_b \leq 6.3$) in the outer-rise and inner-trench regions is observed between 1987 and 1991. A similar distribution of seismicity in the outer-rise and inner-trench regions is observed with earthquakes ($m_b < 5.5$). In northern Chile there is a relative absence of compressional activity ($m_b \geq 5.5$) near the interplate contact since the sequence of December 21, 1967. After that, only a cluster of low-magnitude compressional events has been located in the area 50 to 100 km from the trench. The compressional activity occurring near the interplate zone in both seismic gaps represents that a seismic preslip is occurring in and near the plate contact. Therefore, if this seismic preslip is associated with the maturity of the gap, the fact that it is larger in southern Peru than in northern Chile may reflect that the former gap is more mature than the latter. However, the more intense downdip tensional activity and the absence of compressional seismicity near the contact zone observed in northern Chile, may also be interpreted as evidence that northern Chile is seismically more mature than southern Peru. Therefore, the observed differences in the distribution of stresses and seismicity analyzed under simple models of stress accumulation and transfer in coupled subduction zones are not sufficient to assess the degree of maturity of a seismic gap.

Key words: Seismicity, spatio-temporal variations, seismic gaps, southern Peru, northern Chile.

Introduction

Southern Peru and northern Chile are two adjacent seismic gaps in the circum-Pacific region (KELLEHER, 1972; MCCANN *et al.*, 1979; NISHENKO, 1985; LAY *et al.*, 1989; COMTE and PARDO, 1991). The last two earthquakes with destructive

---

[1] UACPyP, Instituto de Geofísica, UNAM, México D.F. 04510, México.
[2] Also at Depto de Geología y Geofísica, U. de Chile Casilla 2777, Santiago, Chile.

tsunamis occurred there in 1868 ($M_w = 8.8$, southern Peru) and 1877 ($M_w = 8.8$, northern Chile) (DORBATH *et al.*, 1990; COMTE and PARDO, 1991), and no large thrust events ($M_s > 8.0$) have affected these regions during this century. Thus, both gaps are interesting regions to analyze their seismicity during the last 27 years of reliable hypocenters and focal mechanism catalogs.

Studies of temporal variations of downdip intraplate and outer-rise earthquakes in coupled subduction zones have been reported by several authors (e.g., MAL-GRANGE and MADARIAGA, 1983; CHRISTENSEN and RUFF, 1983, 1988; DMOWSKA *et al.*, 1986, 1988; ASTIZ and KANAMORI, 1986; DMOWSKA and LOVISON, 1988; LAY *et al.*, 1989). These studies show that in some gaps, a precursory combination of shallow compressional outer-rise and tensional downdip events is observed before the occurrence of a great earthquake.

DMOWSKA *et al.* (1988) presented a simple one-dimensional model of stress accumulation and transfer in coupled subduction zones during the earthquake cycle, concluding that the combination of both high compressional stress in the outer-rise and tensional intraplate stress in the descending slab imply that the locked thrust zone is subjected to high shear loading, and consequently are signals of maturity of the corresponding seismic cycle. DMOWSKA *et al.* (1988) also suggested that, in some cases, the lower portions of the interplate thrust contact undergo periods of active seismicity, mainly in the last part of the cycle.

These periods of activity are preceded by a temporary quiescence of extensional seismicity downdip in the slab, suggesting that thrust events in the lower, and hence hotter, regions of the contact zone may be a sign of an aseismic preslip taking place there prior to a great earthquake. However, if a preslip episode is included in their model, the behavior in the downdip and the outer-rise zones is appreciably modified; if this preslip is large enough, it can reverse the predicted shallow compressional outer-rise and tensional downdip events prior to the expected great earthquake. Thus a combination of tensional outer-rise earthquakes and compressional downdip events could also be a precursory signal of terminal maturity in a subduction zone, resulting in an ambiguous signal of the stress transfer model.

LAY *et al.* (1989) presented a systematic analysis of circum-Pacific subduction zones where large interplate thrust events occur. They conclude that outer-rise compressional events have occurred prior to several large thrust events, whereas outer-rise tensional events generally occur after interplate ruptures. In the intermediate depth range, large downdip tensional events generally precede interplate thrust faulting, and are often concentrated near the downdip edge of the coupled zone. LAY *et al.* (1989) also proposed a schematic dynamic stress model that summarizes their observations.

Taking into account that the southern Peru and northern Chile seismic gaps are considered mature seismic gaps (e.g., LAY *et al.*, 1989; COMTE and PARDO, 1991), the aim of this work is to analyze the space-time variations of the seismicity in both gaps, in order to see whether they present the predicted precursory combination observed in other seismic gaps.

Teleseismic Data used in the Analysis

The presumed rupture areas of the 1868 and 1877 great earthquakes are presented in Figure 1. Unfortunately, no permanent local seismic networks exist in southern Peru and northern Chile. Thus only a spatio-temporal analysis of the seismicity observed with global networks can be carried out. In this analysis, the USGS-NEIC (1965–1988) and PDE (1989–1991) catalogs are used and a cut-off magnitude of $m_b = 5.0$ was determined for the whole region based on the $\mathrm{Log}(N)$ versus m_b relation (Figure 1). However, this analysis is done only for events of magnitudes greater than 5.5 because they present a better epicentral control and the majority of them have well constrained focal mechanism solutions.

For southern Peru and northern Chile regions, a plot of distances normal to the trench versus time was drawn in order to observe globally the spatio-temporal

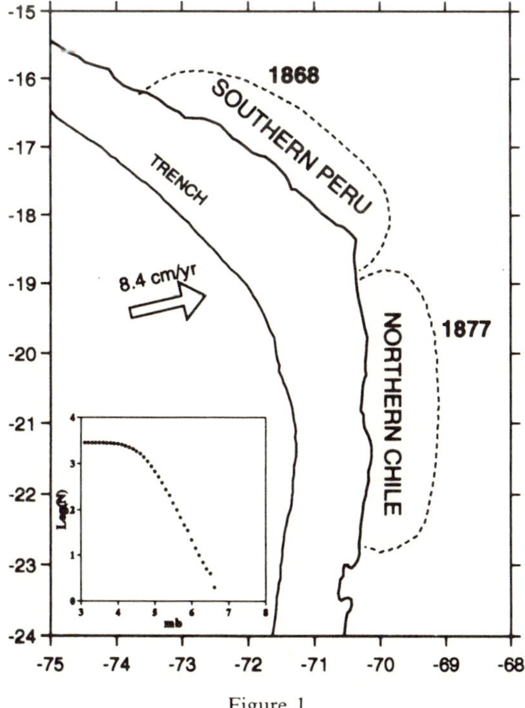

Figure 1

Southern Peru and northern Chile regions, where the dashed lines indicate the 1868 and 1877 estimated rupture zones (COMTE and PARDO, 1991). Earthquake frequency-magnitude relation for southern Peru and northern Chile regions using the NEIC (1965–1988) and PDE (1989–1991) catalogs is also shown in the lower-left corner. N is the accumulated number of events of magnitude m_b or greater.

behavior of the seismicity including the outer-rise and the downdip regions (Figures 2a, 2b and 3a). Table 1 presents the events with reported focal mechanisms used in the following discussion.

For the region located between distances of -100 to 100 km from the trench, an analysis of lower magnitude events ($5.0 \leq m_b < 5.5$) is also carried out in order to observe in more detail the variations of seismicity in this specific region (Figures 2c and 3b). This is because the outer-rise region exhibits the least ambiguous behavior, where compressional activity is observed apparently only prior to major thrust earthquakes (CHRISTENSEN and RUFF, 1988; LAY et al., 1989).

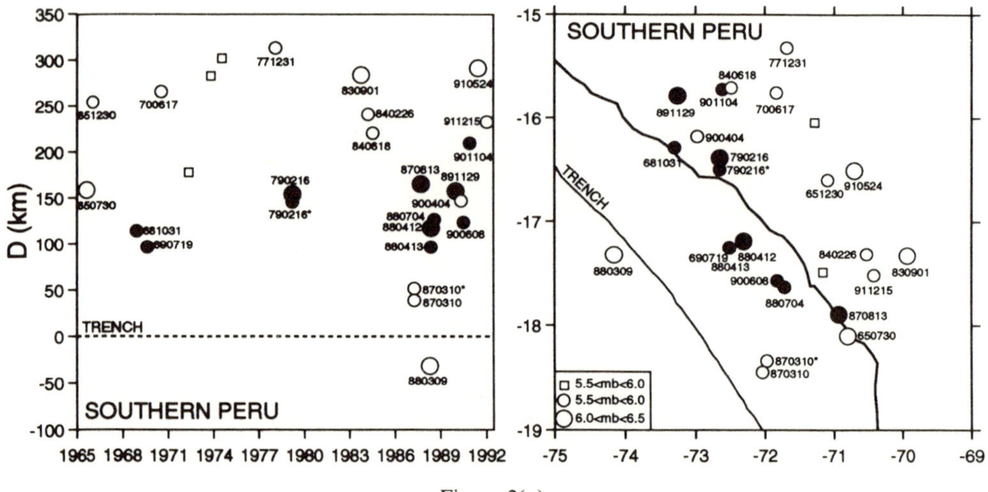

Figure 2(a)

Figure 2(b)

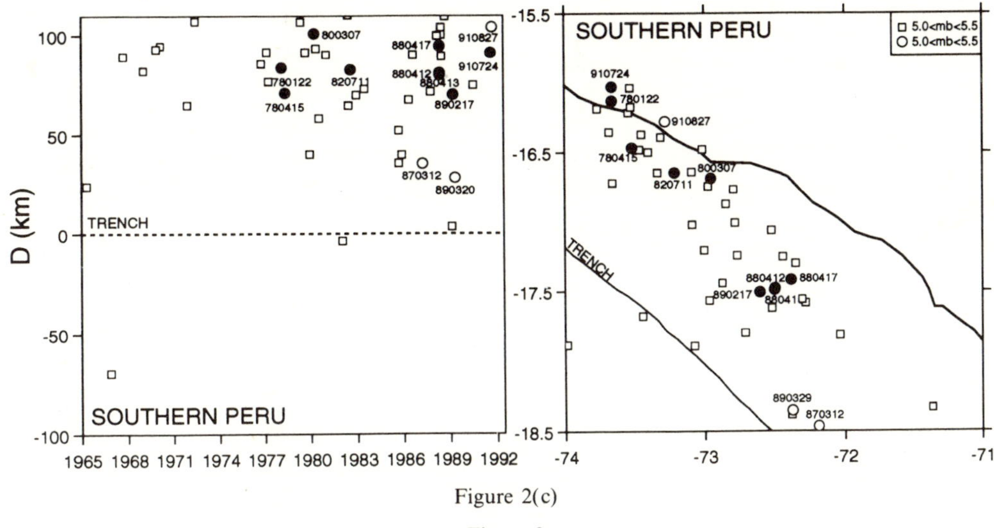

Figure 2(c)

Figure 2

(a) To the left, the distance from the trench (D, km) versus time (yr) plot of events ($m_b \geq 5.5$) in southern Peru is shown. The dashed line indicates the location of the trench. To the right, the epicentral distribution of events is presented. Solid circles denote compressional events, open circles denote normal faulting, and open rectangles correspond to unknown focal mechanism solutions (Table 1). The size of the symbols is related to the magnitude of the events, as is indicated on the Figure. Each event with a known focal mechanism solution is identified with the year, month and day of occurrence; an asterisk is used for the second event, when two earthquakes occurred on the same day. (b) Enlargement of the data shown in Figure 2a for the 1987–1991 time interval. Symbols as in Figure 2a. (c) Distance from the trench versus time plot (to the left) and epicentral distribution (to the right) of events $5.0 \leq m_b < 5.5$, in the inner- and outer-rise southern Peru region.

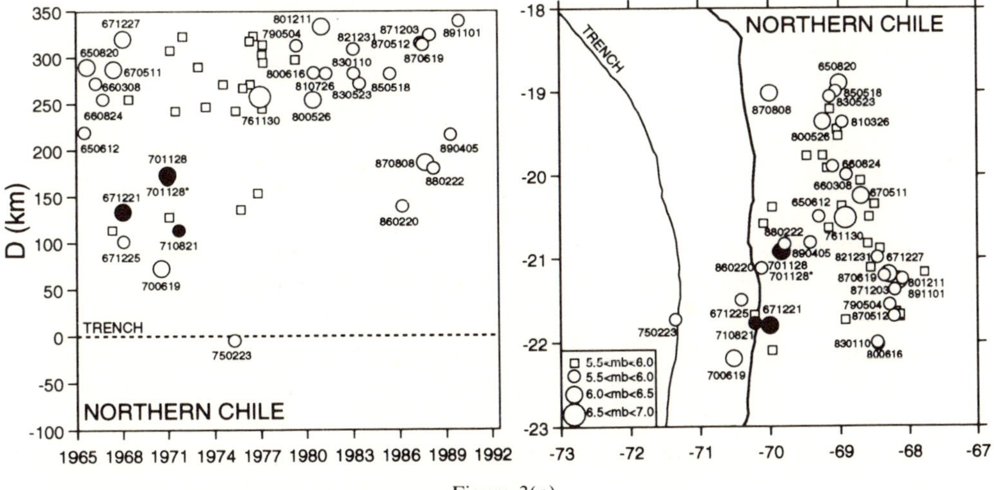

Figure 3(a)

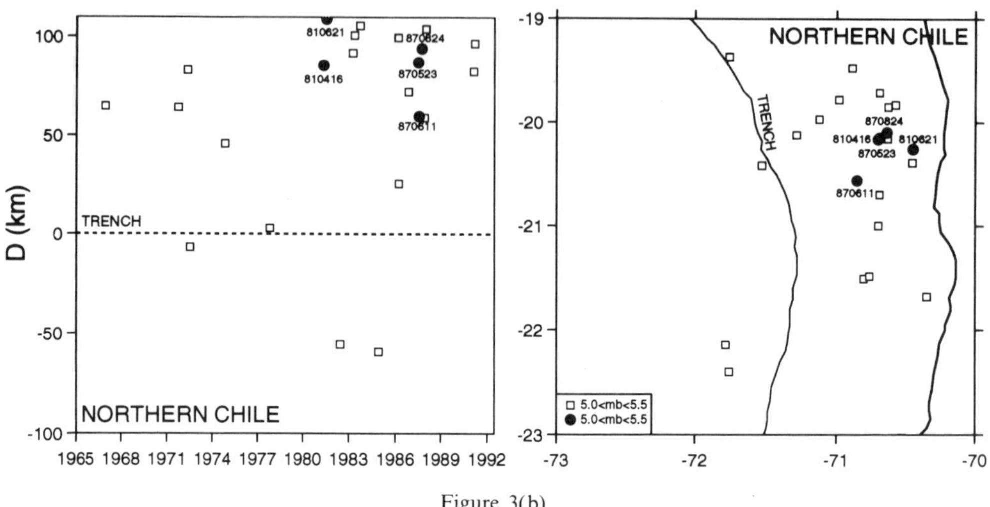

Figure 3(b)

Figure 3

(a) Distance from the trench versus time plot (to the left) and epicentral distribution (to the right) of events $m_b \geq 5.5$ in northern Chile. Symbols as in Figure 2a. (b) Distance from the trench versus time plot (to the left) and epicentral distribution (to the right) of events $5.0 \leq m_b < 5.5$, in the inner- and outer-rise northern Chile region. Symbols as in Figure 2a.

Spatio-temporal Variation of Seismicity in Southern Peru

In southern Peru, the sequence of great historical earthquakes is well documented, with at least three great events reported in 1604, 1784 and 1868 with estimated magnitudes greater than 8.5, determined from macroseismic information (DORBATH *et al.*, 1990; COMTE and PARDO, 1991). Between the 1604 and the 1784 earthquakes, two other large events occurred in 1687 and 1715, with estimated magnitudes of between 7.5 and 8.0; together, both events apparently filled the estimated rupture zone of the great events that occur in the region. Based on this sequence of great earthquakes, it is reasonable to expect that this region is mature and nearing the next great event (COMTE and PARDO, 1991).

If this were the case, the expected behavior of the seismicity, as was observed in other subduction zones, would consist of shallow compressional earthquakes near the outer-rise and tensional downdip events at depth. This stress distribution has been interpreted as a result of a locked interplate boundary where the oceanic plate in the outer-rise is under increased compression, and the downgoing slab stretches under tension from the interplate coupled zone.

Four compressional events ($m_b \geq 6.0$) can be observed in southern Peru (Figure 2a): on November 29, 1989 in the northern edge of the presumed 1868 rupture area,

Table 1

Catalog of events with focal mechanism solutions used in this study. Yr M D: Date (year, month, day); h:m: Origin time (hour : minute); T: N (tensional) and C (compressional). References: ST73: STAUDER (1973); IM71 ISACKS and MOLNAR (1971); ST75: STAUDER (1975); AL88: ASTIZ et al. (1988); AR92: ARAUJO and SUÁREZ (1993); CMT: Centroid Moment Tensor (Harvard); CS: this work

Yr M D h:m	Lat (S)	Lon (W)	Depth	m_b	T	Ref.
1965 06 12 18:50	−20.500	−69.300	102	5.8	N	ST73
1965 07 30 05:45	−18.100	−70.800	72	6.0	N	ST73
1965 08 20 09:42	−18.900	−69.000	128	6.2	N	ST73
1965 12 30 06:16	−16.600	−71.100	112	5.7	N	IM71
1966 03 08 20:46	−20.000	−68.900	112	5.7	N	ST73
1966 08 24 07:17	−19.900	−69.100	100	5.5	N	ST73
1967 05 11 15:05	−20.258	−68.691	79	6.1	N	ST73
1967 12 21 02:25	−21.800	−70.000	33	6.3	C	ST73
1967 12 25 10:41	−21.500	−70.400	53	5.8	N	ST73
1967 12 27 09:17	−21.200	−68.300	135	6.4	N	ST73
1968 10 31 09:15	−16.289	−73.298	67	5.7	C	ST75
1969 07 19 04:54	−17.254	−72.519	54	5.9	C	ST75
1970 06 17 04:44	15.753	71.835	91	5.9	N	ST75
1970 06 19 10:56	−22.193	−70.515	52	6.2	N	ST73
1970 11 28 11:08	−20.921	−69.830	33	6.0	C	ST73
(*) 1970 11 28 14:45	20.924	−69.872	34	5.9	C	CS
1971 08 14 15:00	−21.775	−67.230	189	5.7	N	AR92
1971 08 21 12:47	−21.774	−70.211	17	5.8	C	CS
1975 02 23 03:53	−21.727	−71.356	33	5.6	N	CS
1976 11 30 00:40	−20.520	−68.919	82	6.5	N	AL88
1977 12 31 07:53	−15.300	−71.680	158	5.9	N	CMT
1978 01 22 21:19	−16.131	−73.656	49	5.4	C	CMT
1978 04 15 13:49	−16.470	−73.514	50	5.3	C	CMT
1979 02 16 10:08	−16.390	−72.658	53	6.2	C	CMT
(*) 1979 02 16 22:18	−16.500	−72.656	52	5.5	C	CMT
1979 05 04 17:21	−21.560	−68.284	81	5.6	N	CMT
1979 05 21 22:22	−15.250	−70.089	208	6.0	C	CMT
1979 09 15 10:02	−15.634	−69.559	214	5.6	C	CMT
1980 03 07 08:25	−16.689	−72.952	43	5.4	C	CMT
1980 05 26 18:41	−19.364	−69.238	114	6.1	N	CMT
1980 06 16 05:45	−22.029	−68.459	87	5.5	N	CMT
1980 12 11 18:15	−21.272	−68.153	80	6.1	N	CMT
1981 03 26 18:04	−19.370	−68.957	138	5.8	N	CMT
1981 04 16 22:05	−20.162	−70.699	33	5.1	C	CMT
1981 06 21 10:30	−20.256	−70.446	36	5.2	C	CMT
1982 07 11 02:13	−16.649	−73.214	38	5.3	C	CMT
1982 09 15 20:22	−14.493	−70.785	128	6.0	N	CMT
1982 12 31 03:47	−20.993	−68.464	118	5.7	N	CMT
1983 01 10 09:17	−22.006	−68.470	121	5.6	N	CMT
1983 05 23 00:54	−19.065	−69.139	110	5.5	N	CMT
1983 09 01 20:01	−17.330	−69.932	105	6.0	N	CMT
1984 02 26 08:18	−17.316	−70.526	113	5.8	N	CMT
1984 06 18 11:20	−15.705	−72.491	116	5.8	N	CMT

D. Comte and G. Suárez

Table 1 (*Contd*)

Yr M D h:m	Lat (S)	Lon (W)	Depth	m_b	T	Ref.
1985 05 18 16:59	−19.000	−69.053	109	5.5	N	CMT
1986 02 20 09:16	−21.122	−70.116	33	5.7	N	CMT
1986 03 15 11:29	−18.909	−67.391	243	5.6	N	CMT
1987 03 10 00:22	−18.448	−72.035	41	5.7	N	CMT
(*) 1987 03 10 02:18	−18.341	−71.976	36	5.5	N	CMT
1987 03 12 06:07	−18.467	−72.192	35	5.1	N	CMT
1987 05 12 16:12	−21.694	−68.222	77	5.5	N	CMT
1987 05 23 05:01	−20.151	−70.688	36	5.3	C	CMT
1987 06 11 05:08	−20.559	−70.855	33	5.1	C	CMT
1987 06 19 19:00	−21.213	−68.362	86	5.6	N	CMT
1987 08 08 15:48	−19.022	−69.991	69	6.4	N	CMT
1987 08 13 15:23	−17.897	−70.931	36	6.1	C	CMT
1987 08 24 06:09	−20.096	−70.632	33	5.2	C	CMT
1987 12 03 11:04	−21.380	−68.215	117	5.5	N	CMT
1988 02 22 19:13	−20.833	−69.785	70	5.9	N	CMT
1988 03 09 21:33	−17.327	−74.154	32	6.0	N	CMT
1988 04 12 23:19	−17.192	−72.305	33	6.1	C	CMT
1988 04 12 23:55	−17.478	−72.503	33	5.4	C	CMT
1988 04 13 00:39	−17.256	−72.518	16	5.9	C	CMT
1988 04 13 06:22	−17.489	−72.505	36	5.2	C	CMT
1988 04 17 02:50	−17.420	−72.387	33	5.4	C	CMT
1988 07 04 13:54	−17.636	−71.718	19	5.8	C	CMT
1989 02 17 15:12	−17.509	−72.609	48	5.3	C	CMT
1989 03 29 04:06	−18.353	−72.380	42	5.4	N	CMT
1989 04 05 23:47	−20.816	−69.427	115	5.6	N	CMT
1989 11 01 06:40	−21.252	−68.099	143	5.8	N	CMT
1989 11 29 01:00	−15.781	−73.254	74	6.1	C	CMT
1990 01 05 13:03	−19.258	−69.529	109	5.0	N	CMT
1990 04 04 05:47	−16.179	−72.974	93	5.5	N	CMT
1990 04 20 18:23	−14.859	−71.374	144	5.1	N	CMT
1990 06 08 23:49	−17.573	−71.824	27	5.6	C	CMT
1990 06 18 19:02	−21.825	−68.438	117	5.3	N	CMT
1990 06 21 11:48	−19.648	−69.128	103	5.3	N	CMT
1990 08 15 13:23	−18.834	−69.096	120	5.2	N	CMT
1990 08 28 08:58	−19.645	−69.876	64	5.2	N	CMT
1990 10 10 01:00	−19.503	−66.618	266	5.8	N	CMT
1990 11 04 18:13	−15.721	−72.619	121	5.4	C	CMT
1991 05 24 20:50	−16.506	−70.701	128	6.3	N	CMT
1991 07 09 05:54	−20.599	−68.803	101	5.3	N	CMT
1991 07 23 19:44	−15.679	−71.574	5	5.0	N	CMT
1991 08 27 04:16	−20.855	−68.775	123	5.0	N	CMT
1991 08 27 11:46	−16.278	−73.278	74	5.3	N	CMT
1991 08 27 11:58	−21.657	−68.506	116	5.1	N	CMT
1991 08 27 11:46	−16.278	−73.278	74	5.3	N	CMT
1991 12 02 17:27	−15.867	−69.288	241	5.1	N	CMT
1991 12 15 18:56	−17.521	−70.422	104	5.6	N	CMT

on August 13, 1987 in the southern edge of the 1868 rupture area, and the two subsequent events located around the middle of that area, on February 16, 1979 and April 12, 1988. The presence of frequent tensional intraplate activity can also be observed in southern Peru (Figure 2a), and a change from a compressional to a tensional regime can be noted at about 160 km from the trench; this change is probably associated with the coupled-uncoupled transition of the interplate contact zone. That is, from distances of about 160 to 350 km from the trench the seismic activity is mainly tensional, and from 100 to 160 km from the trench the seismicity is mainly compressional.

It is interesting to note that the two events: the normal-faulting event of July 30, 1965 (72 km depth) and the thrust event of August 13, 1987 (36 km depth) occurred at approximately the same epicenter but with a difference in depth of about 36 km. This is similar to the sequence of the normal-faulting event of March 3, 1965 (72 km depth) and the thrust-faulting event of July 9, 1971 (52 km depth) that occurred near La Ligua (central Chile) (MALGRANGE and MADARIAGA, 1983). In this case, with a longer time interval between them (about 22 years) than in La Ligua.

Offshore, an absence of seismicity between -100 to 100 km from the trench can be observed between 1965 and 1987, followed by two tensional events that occurred on March 10, 1987 ($m_b = 5.5, 5.7$) at about 50 km from the trench, preceding a larger normal faulting outer-rise event on March 9, 1988 ($m_b = 6.0$) (Figure 2a). The reverse and normal faulting events that occurred after 1987 between -100 to 350 km from the trench (Figure 2b) may be explained by a mechanical pull and push interaction along the slab.

The compressional events near the contact zone that have occurred since 1987 in southern Peru, could be interpreted as a sign of a seismic preslip near the seismogenic contact zone. This apparent preslip and the burst of tensional intraplate activity may reflect the maturity of the southern Peru seismic gap. However, it does not present the expected precursory compressional outer-rise events in the case of earthquakes with magnitudes $m_b \geq 5.5$.

The behavior observed in the outer-rise and the inner-trench regions for low magnitude events ($5.0 \leq m_b < 5.5$) is very similar to that observed for earthquakes of greater magnitude ($m_b \geq 5.5$); tensional events are observed near the trench and compressional events downdip from it (Figure 2c). There are four low magnitude ($5.0 \leq m_b < 5.5$) compressional events in southern Peru on January 22, 1978 ($m_b = 5.4$), April 15, 1978 ($m_b = 5.3$), March 7, 1980 ($m_b = 5.4$) and July 11, 1982 ($m_b = 5.3$) which are located close to the compressional event of February 16, 1979 ($m_b = 6.2$) and its main aftershock ($m_b = 5.5$). This compressional sequence between 1978 and 1982 preceded by about nine years the one that occurred between 1987 and 1989. The two outer-rise events and the two events located very close to the trench are too small ($m_b \sim 5.0$) to determine a CMT solution or a first motion focal mechanism.

Spatio-temporal Variation of Seismicity in Northern Chile

The distribution of seismicity in northern Chile is more complicated than that of southern Peru because the sequence of historical earthquakes is not well established. The last great earthquake occurred in 1877. Unfortunately, the preceding great events are not well documented due to the absence of population before the XIX century. This region is in a similar tectonic environment as that of southern Peru, so the periods of recurrence in both regions are probably of the same order. Thus northern Chile has been considered, as has the southern Peru region, as a seismic gap of substantial maturity (LAY et al., 1989; COMTE and PARDO, 1991).

Only two compressional events ($m_b \geq 6.0$) have occurred in northern Chile (Figure 3a) since 1965: the December 21, 1967 thrust event ($m_b = 6.3$) located near the southern edge of the presumed rupture area of the 1877 great earthquake, and the November 28, 1970, compressional event ($m_b = 6.0$) located at about 100 km to the north of the 1967 earthquake. Large downdip tensional events ($m_b \geq 6.0$) in the subducted slab exhibit a quasi-random distribution in time and space. The largest tensional earthquakes are those of August 20, 1965 ($m_b = 6.2$), May 11, 1967 ($m_b = 6.1$), December 27, 1967 ($m_b = 6.4$), June 19, 1970 ($m_b = 6.2$), November 30, 1976 ($m_b = 6.5$), May 26, 1980 ($m_b = 6.1$), December 12, 1980 ($m_b = 6.1$), and August 8, 1987 ($m_b = 6.4$).

MALGRANGE and MADARIAGA (1983) studied a sequence of earthquakes near Tocopilla (21.5°S) that started with a thrust event on December 21, 1967 ($m_b = 6.4$). This interplate event was followed by a sequence of deeper, intraplate earthquakes in the downgoing slab showing a combination of tensional (December 25, 1967: $m_b = 5.8$, and June 19, 1970: $m_b = 6.2$), and compressional (November 28, 1970: $m_b = 6.0$) events. This suite of events occurred in close epicentral proximity. Without direct evidence of a double seismic zone in northern Chile, MALGRANGE and MADARIAGA (1983) interpreted the tensional events as a possible effect of the slab pull felt below the seismogenic zone associated with the December 21, 1967 thrust event.

In order to complete the sequence of the 1967–1971 earthquakes, the focal mechanisms of the August 21, 1971 ($m_b = 5.5$) event and the November 28, 1970 ($m_b = 5.9$) aftershock were determined using the polarity of first motions reported by the ISC bulletin (Figures 4a and 4b). Although the nodal planes of the fault plane solutions are not constrained, it is clear from the first motion data that they are reverse faulting events. As in southern Peru, the change from compressional to tensional regimes along the slab in northern Chile occurs at a distance of about 160 km from the trench (Figure 3a). The presence of frequent tensional activity in the subducted slab observed in the northern Chile gap may be interpreted as the result of the strongly coupled seismogenic thrust zone associated with the 1877 earthquake.

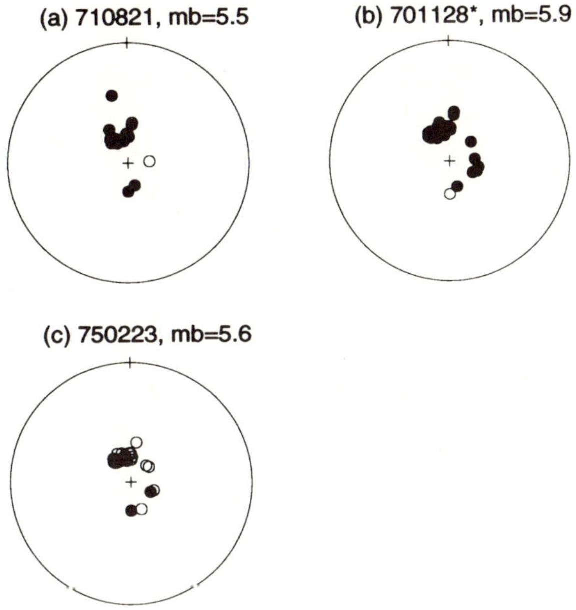

Figure 4

Lower hemispheric projections of the first motion polarities reported by the ISC bulletin for three northern Chile events. Closed and open circles represent compressional and dilational first motion. The data are not sufficient to constrain the fault planes, but it can be observed that the August 21, 1971 ($m_b = 5.5$) and the November 28, 1970 ($m_b = 5.9$) earthquakes are compressional; and the February 23, 1975 ($m_b = 5.6$) outer-rise event is tensional.

Northern Chile does not show the presence of outer-rise seismic activity ($m_b \geq 6.0$). In fact, there is only one outer-rise event which occurred on February 23, 1975 ($m_b = 5.6$) (Figure 3a). The focal mechanism of this event was determined based on the first motion reported by the ISC and it shows normal faulting. Unfortunately, the data is not sufficient to constrain the fault planes (Figure 4c). As was shown for southern Peru, the northern Chile gap also does not present compressional outer-rise events of $m_b \geq 5.5$. However, four lower magnitude, outer-rise events ($5.0 \leq m_b < 5.5$) can be observed in northern Chile (Figure 3b); no reliable focal mechanisms are available for these events. The behavior of the compressional seismicity (Figure 3b) for this lower magnitude interval, is different to that observed for higher magnitudes ($m_b \geq 5.5$). The absence of seismicity observed at distances of 0 to about 100 km from the trench for events $m_b \geq 5.5$ is filled with smaller events $5.0 \leq m_b < 5.5$. A cluster of compressional events can be observed between 50 to 100 km from the trench, from May to August, 1987. This distribution of seismicity may be interpreted as seismic preslip occurring near the coupled zone, in a manner similar to that observed in southern Peru, but with lower magnitude events.

Summary and Conclusions

The purpose of this work was to analyze two clearly recognized seismic gaps, in southern Peru and northern Chile, making use of current models of space-time variations in seismicity in order to elucidate their degree of maturity. Both gaps present some similarities: (1) the last great earthquake in each region occurred about 120 years ago, and no large thrust events have taken place within the gaps during the present century; (2) in both cases, the rate of convergence and the age of the oceanic sea-floor near the trench are almost identical. Therefore, it is expected that they would exhibit very similar patterns of seismicity. Although some similarities exist, the distribution of seismicity in the two gaps presents some important differences, mainly near the interplate contact zone. At this point, it is interesting to mention that, although observations in other subduction zones of the world show that earthquake recurrence times and rupture length can vary between successive earthquake cycles (e.g., COMTE *et al.*, 1986; THATCHER, 1990; RUFF, 1992), the fact that the last great earthquakes that affected southern Peru (1868) and northern Chile (1877) have rupture lengths of about 450 km, suggests that they could represent the largest events of these regions. Therefore, the next great earthquakes may not necessarily be of the same magnitude as the latest events, but probably they will occur within the seismic gap areas analyzed by this work.

Neither southern Peru nor the northern Chile gaps clearly exhibit the precursory compressional outer-rise events during the time period studied. Furthermore, both gaps present downdip intraplate tensional seismicity in the subducted slab, which is more intense in the northern Chile segment. For both regions, the change from compressional to tensional regime within the subducted slab occurs at about 160 km from the trench, which is apparently associated with the coupled-uncoupled transition of the interplate contact zone.

There are, however, some differences in the space-time distribution of stresses and of seismicity in both gaps. In southern Peru, an increase of compressional seismicity near the contact zone is observed since 1987. In contrast, the northern Chile gap exhibits an absence of compressional activity ($m_b \geq 5.5$) near the interplate contact since the December 21, 1967 sequence. This absence, however, is apparently filled with a cluster of lower magnitude compressional events that occurred during 1987. Also, the southern Peru gap shows a burst of tensional activity near the inner- and outer-rise zone during 1987, which is not observed in northern Chile.

Under an interpretation of stress transfer in the subduction zones, the continuous downdip tensional activity observed in southern Peru and norhtern Chile may be interpreted as evidence that these regions are strongly coupled. However, the fact that both the downdip tensional activity is more frequent in northern Chile than in southern Peru and that there is an absence of compressional activity ($m_b \geq 5.5$) near the contact zone in northern Chile, may be considered as evidence that this gap is more mature than southern Peru.

On the other hand, the burst of both compressional seismicity near the contact zone, and the tensional activity near the inner- and outer-rise regions observed in southern Peru since 1987, may be interpreted in turn as a sign that a seismic preslip is taking place near the interplate contact. If the apparent seismic preslip is associated with the maturity of the gap, the fact that it is observed in northern Chile with lower magnitude earthquakes than those observed in southern Peru, may then be interpreted as representing that southern Peru is a more mature seismic gap than northern Chile.

In summary, contradictory results are obtained about the maturity of the northern Chile and southern Peru gaps using alternative interpretations of simple stress distribution and transfer models in and near the coupled interface in a subduction zone. Although similarities are observed in both gaps, there are sharp differences in the spatial distribution of seismicity and in the stress patterns in both segments of this subduction zone.

These results suggest that the complexity of the observed distribution of seismicity in seismic gaps cannot always be explained with simple models of accumulation and transfer of stresses along the slab. Futhermore, the available database (~ 30 years) of reliably located seismicity for which focal mechanisms exist may not be sufficiently long to draw firm conclusions about the long-term distribution of stresses near the coupled zone.

Acknowledgements

This work was greatly improved by the comments of R. Dmowska and one anonymous reviewer. We thank S. K. Singh and D. Byrne for their suggestions. The research was partially funded by Fundación Andes, Chile grant C–52040.

REFERENCES

ARAUJO, M., and SUÁREZ, G. (1993), *Detailed Geometry and State of Stress of the Subducted Nazca Plate beneath Central Chile and Argentina: Evidence from Teleseismic Focal Depth Determinations*, Geophys. J. Int., in press.

ASTIZ, L., and KANAMORI, H. (1986), *Interplate Coupling and Temporal Variation of Mechanisms of Intermediate-depth Earthquakes in Chile*, Bull. Seismol. Soc. Am. *76*, 1614–1622.

ASTIZ, L., LAY, T., and KANAMORI, H. (1988), *Large Intermediate-depth Earthquakes and the Subduction Process*, Phys. Earth Planet. Inter. *53*, 80–166.

CHRISTENSEN, D., and RUFF, L. (1983), *Outer-rise Earthquakes and Seismic Coupling*, Geophys. Res. Lett. *10*, 697–700.

CHRISTENSEN, D., and RUFF, L. (1988), *Seismic Coupling and Outer-rise Earthquakes*, J. Geophys. Res. *93*, 13421–13444.

COMTE, D., EISENBERG, A., LORCA, E., PARDO, M., PONCE, L., SARAGONI, R., SINGH, S. K., and SUÁREZ, G. (1986), *The 1985 Central Chile Earthquake: A Repeat of Previous Great Earthquakes in the Region?*, Science *233*, 449–453.

COMTE, D., and PARDO, M. (1991), *Reappraisal of Great Historical Earthquakes in the Northern Chile and Southern Peru Seismic Gaps*, Natural Hazards *4*, 23–44.

DMOWSKA, R., LOVISON, L., and RICE, J. (1986), *Stress Accumulation and Transfer between a Subduction Slab and the Thrust Contact Zone during the Whole Earthquake Cycle* (Abstract), EOS Trans. AGU *67*, 307.

DMOWSKA, R., RICE, J., LOVISON, L., and JOSSELL, D. (1988), *Stress Transfer and Seismic Phenomena in Coupled Subduction Zones during the Earthquake Cycle*, J. Geophys. Res. *93*, 7869–7884.

DMOWSKA, R., and LOVISON, L. (1988), *Intermediate-term Seismic Precursors for Some Coupled Subduction Zones*, Pure and Appl. Geophys. *126*, 643–664.

DORBATH, L., CISTERNAS, A., and DORBATH, C. (1990), *Quantitative Assessment of Great Earthquakes in Peru*, Bull. Seismol. Soc. Am. *80*, 551–576.

ISACKS, B., and MOLNAR, P. (1971), *Distribution of Stresses in the Descending Lithosphere from a Global Survey of Focal-mechanism Solutions of Mantle Eathquakes*, Rev. Geophys. Space Phys. *9*, 103–174.

KELLEHER, J. A. (1972), *Rupture Zones of Large South American Earthquakes and Some Predictions*, J. Geophys. Res. *77*, 2087–2103.

LAY, T., ASTIZ, L., KANAMORI, H., and CHRISTENSEN, D. (1989), *Temporal Variation of Large Intraplate Earthquakes in Coupled Subduction Zones*, Phys. Earth Planet. Inter. *54*, 258–312.

MALGRANGE, M., DESCHAMPS, A., and MADARIAGA, R. (1981), *Thrust and Extensional Faulting under the Chilean Coast: 1965, 1971 Aconcagua Earthquakes*, Geophys. J. R. Astron. Soc. *66*, 313–331.

MALGRANGE, M., and MADARIAGA, R. (1983), *Complex Distribution of Large Thrust and Normal-fault Earthquakes in the Chilean Subduction Zone*, Geophys. J. R. Astron. Soc. *73*, 489–505.

MCCANN, W., NISHENKO, S., SYKES, L., and KRAUSE, J. (1979), *Seismic Gaps and Plate Tectonics: Seismic Potential for Major Boundaries*, Pure and Appl. Geophys. *117*, 1082–1147.

NISHENKO, S. (1985), *Seismic Potential for Large and Great Interplate Earthquakes along the Chilean and Southern Peruvian Margins of South America. A Quantitative Reappraisal*, J. Geophys. Res. *90*, 3589–3615.

STAUDER, W. (1973), *Mechanism and Spatial Distribution of Chilean Earthquakes with Relation to Subduction of the Oceanic Plate*, J. Geophys. Res. *78*, 5033–5061.

STAUDER, W. (1975), *Subduction of the Nazca Plate under Peru as Evidenced by Focal Mechanisms and by Seismicity*, J. Geophys. Res. *80*, 1053–1064.

RUFF, L. (1992), *Asperity Distributions and Large Earthquake Occurrence in Subduction Zones*, Tectonophys. *211*, 61–83.

THATCHER, W. (1990), *Order and Diversity in the Modes of Circum-Pacific Earthquake Recurrence*, J. Geophys. Res. *95*, 2609–2623.

(Received February 26, 1993, revised May 18, 1993, accepted May 29, 1993)

PAGEOPH, Vol. 140, No. 2 (1993)

0033–4553/93/020331–34$1.50 + 0.20/0

The 1986 Kermadec Earthquake and Its Relation to Plate Segmentation

HEIDI HOUSTON,[1] HELEN ANDERSON,[2] SUSAN L. BECK,[3]
JIAJUN ZHANG,[1] and SUSAN SCHWARTZ[1]

Abstract—To evaluate the tectonic significance of the October 20, 1986 Kermadec earthquake ($M_W = 7.7$), we performed a comprehensive analysis of source parameters using surface waves, body waves, and relocated aftershocks. Amplitude and phase spectra from up to 93 Rayleigh waves were inverted for centroid time, depth, and moment tensor in a two-step algorithm. In some of the inversions, the time function was parameterized to include information from the body-wave time function. The resulting source parameters were stable with respect to variations in the velocity and attenuation models assumed, the parameterization of the time function, and the set of Rayleigh waves included. The surface wave focal mechanism derived ($\phi = 275°$, $\delta = 61°$, $\lambda = 156°$) is an oblique-compressional mechanism that is not easy to interpret in terms of subduction tectonics. A seismic moment of 4.5×10^{20} N-m, a centroid depth of 45 ± 5 km, and a centroid time of 13 ± 3 s were obtained. Directivity was not resolvable from the surface waves. The short source duration is in significant contrast to many large earthquakes.

We performed a simultaneous inversion of P and SH body waves for focal mechanism and time function. The focal mechanism agreed roughly with the surface wave mechanism. Multiple focal mechanisms remain a possibility, but could not be resolved. The body waves indicate a short duration of slip (15 to 20 s), with secondary moment release 60 s later. Seismically radiated energy was computed from the body-wave source spectrum. The stress drop computed from the seismic energy is about 30 bars. Sixty aftershocks that occurred within three months of the mainshock were relocated using the method of Joint Hypocentral Determination (JHD). Most of the aftershocks have underthrusting focal mechanisms and appear to represent triggered slip on the main thrust interface. The depth, relatively high stress drop, short duration of slip, and paucity of true aftershocks are consistent with intraplate faulting within the downgoing plate. Although it is not clear on which nodal plane slip occurred, several factors favor the roughly E-W trending plane. The event occurred near a major segmentation in the downgoing plate at depth, near a bend in the trench, and near a right-lateral offset of the volcanic arc by 80 km along an E-W direction. Also, all events in the region from 1977 to 1991 with CMT focal mechanisms similar to that of the mainshock occurred near the mainshock epicenter, rather than forming an elongate zone parallel to the trench as did the aftershock activity. We interpret this event as part of the process of segmentation or tearing of the subducting slab. This segmentation appears to be related to the subduction of the Louisville Ridge, which may act as an obstacle to subduction through its buoyancy.

Key words: Tectonics, subduction, plate segmentation, Kermadec, earthquake, rupture process.

---

[1] University of California, Santa Cruz, California, U.S.A.
[2] Institute of Geological and Nuclear Sciences, New Zealand.
[3] University of Arizona, U.S.A.

Introduction

The Kermadec subduction zone has been relatively neglected in the systematic study of subduction zones, perhaps due to its remote location. The October 20, 1986 Kermadec earthquake (06:46:09.98 UT, 28.118°S, 176.367°W, $M_S = 8.1$, National Earthquake Information Center (NEIC)) was the second largest event in 1986. It was anomalous in several respects, including its unusual focal mechanism, relatively high stress drop, and short duration. Additionally, the relationship of the aftershock sequence to the mainshock is not clear. Hence, understanding this event is important for improving the state of our knowledge of earthquake mechanics along the Kermadec subduction zone and in general. The goal of this paper is to evaluate the tectonic significance of the Kermadec earthquake. To this end, we determined source parameters of the 1986 Kermadec earthquake from surface waves, body waves and relocated aftershocks, and considered their implications in the context of the regional tectonics. Additionally, our analysis of this event has implications for issues of resolution in surface and body wave inversions.

The Harvard Centroid Moment Tensor (CMT) focal mechanism for the Kermadec earthquake, as well as the first-motion solution from NEIC (Figure 1), has roughly similar amounts of compressional and strike-slip motion with a highly unusual orientation that is difficult to interpret in terms of subduction of the Pacific plate under the Australian plate. The aftershocks have predominantly underthrusting mechanisms. This raises the possibility that the Kermadec earthquake may have consisted of two or more subevents with different focal mechanisms, each of which

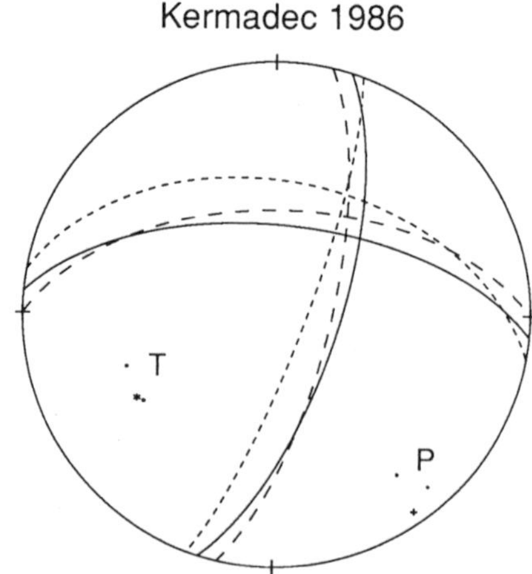

Figure 1
Focal mechanisms (lower hemisphere equal-area projection) of the Kermadec earthquake from *P*-wave first motions (dotted lines), Harvard centroid moment tensor solution (dashed lines), and the moment tensor solution obtained in this study from surface waves (solid lines).

might be more easily understood in terms of the regional tectonic framework. The alternative is that the Kermadec event had a highly unusual focal mechanism, even compared to the relatively small set of large intraplate events in subduction zones. Almost all large, shallow, intraplate events in subduction zones exhibit down-dip tension and can be attributed to slab pull. The largest intraplate events in subduction zones are the 1977 Tonga ($M_W = 8.1$), 1977 Sumba ($M_W = 8.3$), 1970 Peru ($M_W = 7.9$), and 1933 Sanriku ($M = 8.3$) earthquakes. These events have been interpreted as normal-faulting events caused by downdip tension in the subducting slab. Possessing neither downdip tension nor downdip compression, the 1986 Kermadec earthquake cannot be interpreted in this way, and is, in this respect, unique for its size. Therefore, although the Kermadec earthquake has already been studied by LUNDGREN et al. (1989), we felt, in view of the above considerations, that a further, more comprehensive analysis including all available body waves, surface waves, and relocation of aftershocks was warranted. In particular, recent developments in body wave inversions provide algorithms that determine the focal mechanisms of individual subevents (KIKUCHI and KANAMORI, 1991), and recent earthquake studies indicate that some events do consist of subevents with different mechanisms. We utilized this approach and a set of relatively broadband waveforms to evaluate the need for such complexity in modeling the Kermadec earthquake. Our choice of fault plane and our interpretation of the tectonic significance of this event also differ from that of LUNDGREN et al. (1989).

Tectonic Setting

The Kermadec trench is the southern continuation of the Tonga trench; the two trenches are separated at about 26°S by the subducting Louisville Ridge (see Figure 2). This ridge is an elongate chain of seamounts on a broad crustal swell that is similar in many respects to the Hawaiian-Emperor chain, although the current location of the hot spot that produced it is still enigmatic (WATTS et al., 1988; LONSDALE, 1988). The intermediate and deep seismicity in the Tonga-Kermadec region defines the world's most active Benioff zone. The dip of the Benioff-zone seismicity in the Kermadec subduction zone is about 70° at intermediate depths.

RUFF and KANAMORI (1980) characterized the degree of seismic coupling in subduction zones globally, based on an empirical correlation between the size of the largest earthquake in a zone and the convergence rate (positive correlation) or age of the subducting plate (negative correlation). Thus, to first order, convergence rate and age appear to control the strength of mechanical coupling in subduction zones. Using a convergence rate of about 6 cm/yr and an age of 120 million years, Ruff and Kanamori concluded that the Kermadec subduction zone is relatively weakly coupled. The active back-arc opening behind the Tonga and Kermadec trenches is also consistent with weak seismic coupling in this model. Thus, a large component of aseismic slip and relatively few large to great shallow underthrusting earthquakes are expected.

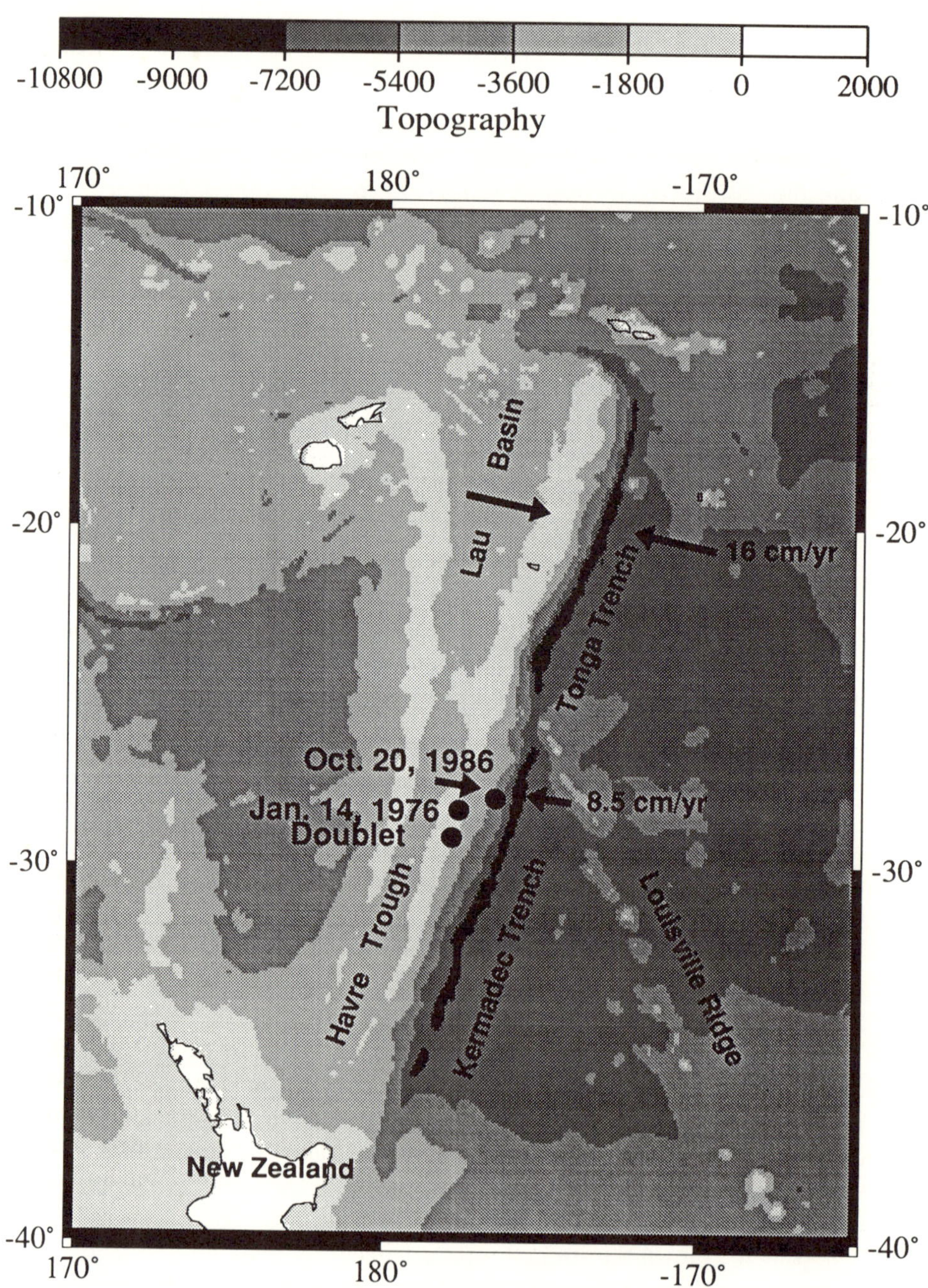

Figure 2

Bathymetric map of the Tonga-Kermadec region. The direction and rate of plate consumption at the trench is indicated by arrows. The epicenters of the 1986 Kermadec earthquake ($M_W = 7.7$) and the 1976 doublet ($M_S = 7.7$ and 8.0) are shown.

The 1986 Kermadec earthquake occurred in a region that has experienced several large earthquakes, the largest of which was the June 22, 1977 normal-faulting event in Tonga with $M_W = 8.2$ (CHRISTENSEN and LAY, 1988). Many of the largest events along the southern section of the Tonga arc appear to be associated with subduction of the Louisville Ridge. The epicenter of the 1986 Kermadec earthquake (Figure 2) lies about 200 km to the south of the intersection of the Louisville Ridge and the Tonga-Kermadec trench. Previous earthquake activity in the Kermadec region includes two large earthquakes with $M_S = 7.7$ and 8.0 separated by only 51 minutes on January 14, 1976. Although determination of the focal mechanism of the second event is complicated by the seismic energy generated by the first event, it is believed that both events of the doublet represent interplate underthrusting (CHRISTENSEN et al., 1991). The 1976 events occurred about 150 km south-southwest of the 1986 event. The aftershock areas of the 1976 doublet and the 1986 earthquake are on adjacent segments along the strike of the trench, but have no significant overlap, based on PDE locations.

The Pacific-Australian convergence direction in this region is oriented approximately east-west. Since the Louisville Ridge is oriented southeast-northwest, its intersection with the Tonga-Kermadec trench is migrating south at an estimated rate of 18 cm/yr (BALLANCE et al., 1989). The Kermadec earthquake is therefore different from the large earthquakes discussed by CHRISTENSEN and LAY (1988) in that it lies in the region where the Louisville Ridge has not yet been subducted.

The region of the 1986 Kermadec earthquake is bounded on the west by the northern Havre Trough, which is actively extending (PELLETIER and LOUAT, 1989). Based on studies of shallow earthquake focal mechanisms, PELLETIER and LOUAT (1989) distinguish the rapidly extending Lau Basin behind the Tonga trench (north of 26°) from the more slowly extending Havre Trough behind the Kermadec trench (south of 26°), and note that the transition between the two regimes corresponds with the arrival at the trench of the Louisville Ridge. The rates of convergence between the Pacific and Indo-Australian plates predicted by the NUVEL-1 model (DEMETS et al., 1987) are 8.1 and 6.7 cm/yr at 20°S and 28°S, respectively. Rates of consumption at the trench are significantly higher than these due to the opening of the Lau Basin (8 cm/yr) and Havre Trough (0.8 to 2.1 cm/yr) behind the arc (PELLETIER and LOUAT, 1989). The rates and directions of the plate consumption shown in Figure 2 were calculated by summing the vectors representing back-arc extension from PELLETIER and LOUAT (1989) and the Pacific-Australian convergence vectors predicted by NUVEL-1.

Aftershock Relocations

The 1986 Kermadec mainshock and sixty aftershocks that occurred between October 20, 1986 and January 4, 1987 with $m_b \geq 5.0$ were relocated using ISC arrival time data and the method of joint hypocentral determination (JHD) described by DEWEY (1972). The first step of this procedure involved the location of 9 well-recorded and well-distributed events with respect to a calibration event, and the simultaneous determination of station corrections for the 100 stations used in the location process. The second step used the station corrections and variances determined by JHD in a single event location program to calculate hypocenters and 90% confidence ellipses and ellipsoids on the epicentral and hypocentral coordinates for all events. We chose the largest Kermadec aftershock (December 21, 1986; $m_b = 6.0$) to be the calibration event and constrained its hypocenter to lie at the ISC epicenter and the Harvard CMT depth. Subsequent body-wave modeling confirmed the CMT depth. The JHD station corrections account for travel-time anomalies that are common to the raypaths of all ten earthquakes to a given station. If the source region is sufficiently small, station corrections should adequately correct for travel time biases resulting from near-source velocity anomalies. Rather than using the JHD method to locate all of the aftershocks, JHD-determined station corrections and variances were used in a single event procedure to prevent problem earthquakes (such as those with foci that located above the surface of the earth) from introducing instabilities into the JHD computation. Locations of the events used to compute station corrections remained unchanged when located with the single event algorithm since the station corrections and weighting functions used in the single event procedure were those computed by JHD.

Fifty-two of the 60 relocated aftershocks had epicenter confidence ellipses and hypocentral ellipsoids of less than 20 km, while 40 events had both epicentral and hypocentral errors of less than 15 km. The relocated aftershocks with errors less than 20 km are shown in map view in a series of time panels in Figure 3. The average change in epicentral location compared with ISC locations was about 10 km, with some tendency for relocated epicenters to move in the downdip direction away from the trench axis, perhaps because the ISC epicenter of the calibration event (the largest aftershock) may be farther west than the real epicenter. This is suggested by the CMT location of this event which is 40 km east of the ISC location.

Several features of the aftershock pattern in Figure 3 are noteworthy. The relocations, like the ISC locations, form a zone elongated 150 km along the strike of the trench. An east-west trending zone may be present near the latitude of the mainshock epicenter. The aftershocks that occurred late in the sequence in November and December 1986 are at the northern and southern ends of the aftershock zone, indicating that an expansion of the aftershock zone took place. Of 33 aftershocks with Harvard CMT focal mechanisms, 26 have underthrusting mechanisms and shallow depths; only one has a mechanism similar to that of the mainshock and that event has a CMT depth of 63 km. The locations and mechanisms of the relocated aftershocks suggest that most of them are not true aftershocks, but rather represent triggered slip on the main underthrusting

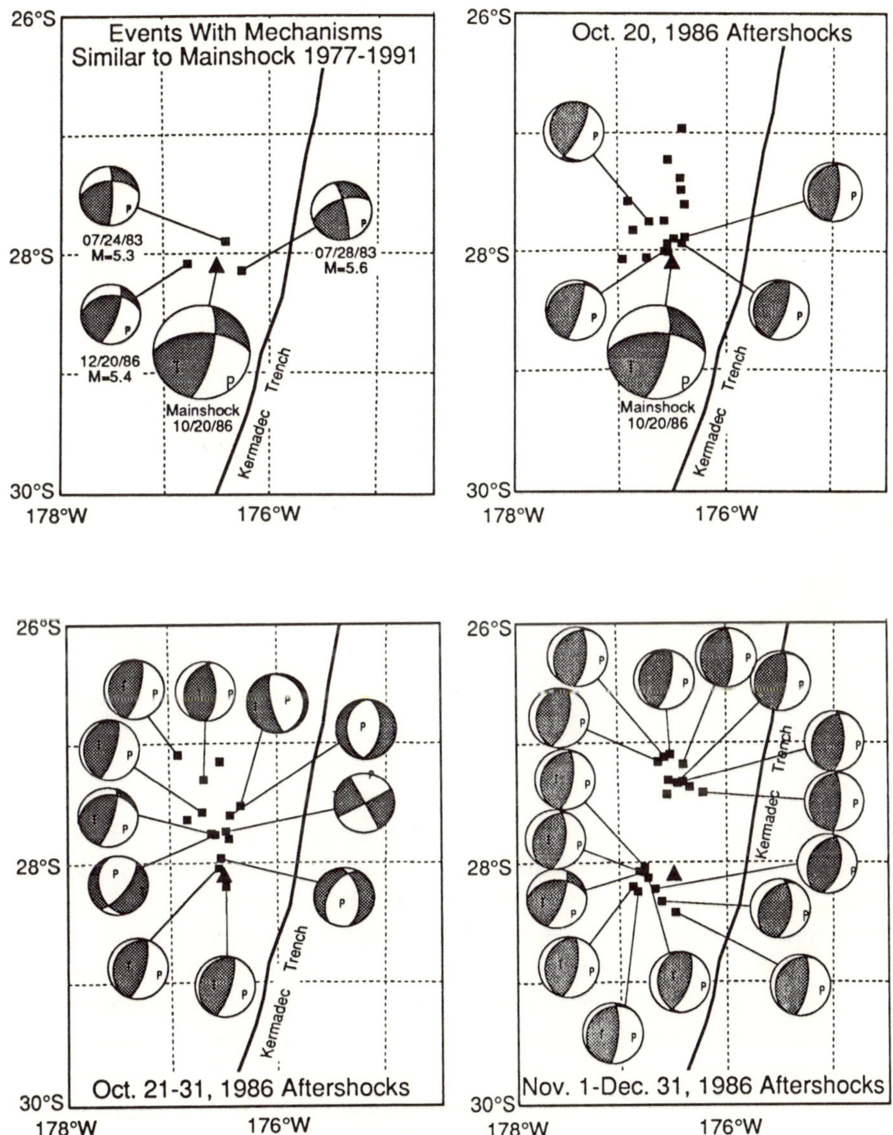

Figure 3

Relocated aftershocks of the 1986 Kermadec earthquake with hypocentral errors less than 20 km. The mainshock mechanism is from the surface wave analysis in this study. Other focal mechanisms are from the CMT catalog. Most of the aftershocks have underthrusting mechanisms. The solid triangle in each panel represents the mainshock epicenter. The heavy line shows the location of the trench taken from the gravity map of DAVY (1989). There is a small change in strike near latitude 28.3°S. The upper left panel shows all events from 1977 to 1991 within the map boundaries that have focal mechanisms similar to that of the mainshock. Note that all three such events are near the mainshock epicenter. Also the depths of the three events are 57, 42, and 62 km, in order of their occurrence. The locations of the 1983 events are from NEIC.

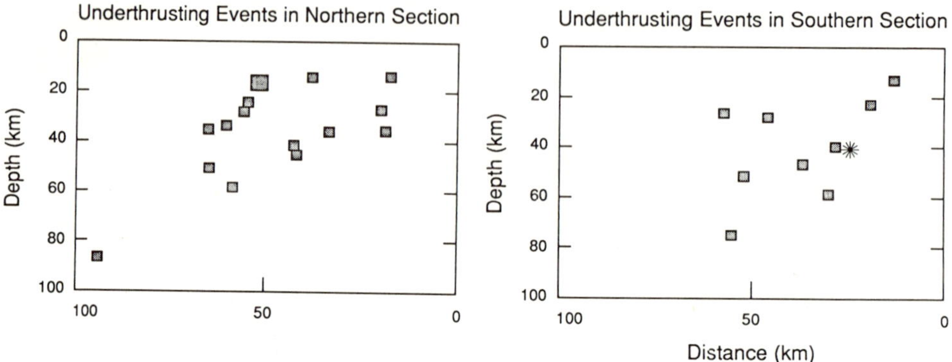

Figure 4

Cross sections of relocated underthrusting aftershocks of the 1986 Kermadec earthquake. The largest aftershock (December 21, 1986) is represented by a larger square. The mainshock location is represented by an asterisk. The southern section appears to be dipping more steeply than the northern section, suggesting segmentation of the downgoing plate.

plate boundary. Waveform modeling (discussed below) for the largest aftershock (December 21, 1986, $M_W = 6.2$) yields a depth of 17 km, consistent with the Harvard CMT depth. The mainshock hypocenter (40 km depth) locates deeper than most of the aftershocks, including the largest aftershock; this suggests that the mainshock faulting occurred within the subducting plate rather than at the plate boundary.

Figure 4 shows cross sections of the Kermadec aftershocks that could be identified as underthrusting events from the CMT catalog, located north and south of 28.046°S (the epicenter of the December 21 aftershock). The mainshock hypocenter is indicated by an asterisk. The trench axis lies 35 km to the east-southeast of the right side of the plots. However, the reader should remember that this depends solely on the ISC location of the calibration event, which may lie farther west than the actual epicenter as mentioned above. The underthrusting earthquakes should be the best located events since they are the larger events and make up the majority of events used to compute JHD station corrections employed in the relocation procedure. Although substantial scatter remains in the locations of the underthrusting aftershocks, as a group the aftershocks in the northern section define a more shallowly-dipping interface than those in the southern section, suggesting that the downgoing plate may be segmented in this region. Weak evidence for such segmentation in this region can also be seen in the PDE catalog.

If the degree to which the relocated hypocenters define a slab in cross-sectional view is a measure of their location accuracy, Figure 4 indicates only fair accuracy at best for our relocated Kermadec aftershocks. A decrease in location accuracy may be expected in the Kermadec arc compared to many other island arcs due to its relatively remote location with respect to seismic stations. Most of the stations used in the relocation were located at teleseismic distances and tended to be concentrated into two azimuthal bands, west-northwest and east-northeast of the aftershock region. Another source of the location error may have arisen from the assumption that the JHD station corrections adequately account for near-source velocity heterogeneity. Although the aftershock region is quite small along the strike of the trench, it contains earthquakes varying in depth between 10 and 100 km. JHD station corrections were computed using events with depths less than 60 km since the majority of the aftershocks occur at shallow depth. These corrections may not be appropriate for the deeper events, since their raypaths may significantly differ from raypaths of the shallower earthquakes used to compute the station corrections.

Surface Wave Analysis

In order to characterize source parameters of the 1986 Kermadec earthquake at long periods, we inverted the surface waves for focal mechanism, depth, duration and moment, following the approach of ZHANG and KANAMORI (1988a). We varied the velocity and attenuation models, the parameterization of the time function, and the set of Rayleigh waves included. Source parameters derived from data of different periods can vary significantly (WALLACE *et al.*, 1991; ANDERSON and ZHANG, 1991). For shallow earthquakes (hypocentral depth <30 km), the duration, centroid depth and moment are particularly sensitive to the period and the earth model used in the inversion. In order to obtain a unified source model for the Kermadec earthquake that is compatible with the body and surface wave data, we used a method similar to that used by ZHANG and LAY (1989a) and ANDERSON and ZHANG (1991). This method incorporates the detailed moment rate function, predetermined from the body waves, in the surface wave inversion.

Systematic bias in the derived source parameters that results from the choice of model for the excitation functions and attenuation is likely to be greater than the formal errors for any specific inversion result, so source parameters are determined as a range of values that correspond to the extremes of the models used.

Data and Inversion Method

We used Rayleigh wave data available from stations of the GDSN, GEO-SCOPE and IDA networks, and 93 R1, R2, and R3 phases were of suitable quality to use.

For our surface wave inversion we use the method of ZHANG and KANAMORI (1988a), which is based on the moment tensor inversion of KANAMORI and GIVEN (1981) modified to a two-step inversion algorithm following ROMANOWICZ and GUILLEMANT (1984). The first step isolates the source finiteness effect (duration), and the second step finds the best depth and moment tensor. Spatial and temporal source finiteness can be determined from surface waves when the earth structure influence on surface wave propagation is known. Because of differences in frequency content, surface waves and free oscillations are useful in resolving the low-frequency component of the source process, while body waves have higher resolution of spatial and temporal finiteness than surface waves.

Our default earth model used in the following analysis incorporates the M84C model of WOODHOUSE and DZIEWONSKI (1984) for the phase velocity, the DZIEWONSKI and STEIM (1982) model for attenuation, and the average ocean model of REGAN and ANDERSON (1984) for calculation of the surface wave excitation functions. These and various other models have been discussed in detail by ZHANG and KANAMORI (1988a). The choice of excitation function model is complicated by the differing ages of the oceanic plates on either side of the Kermadec region. The subducting plate is thought to be Cretaceous in age, whereas the plate to the west is 30–35 Myr (DOUTCH, 1981), so an average ocean model seems appropriate.

Simple Point Source

We assumed a boxcar source time function and performed a two-step inversion of surface wave spectra using two attenuation models (our default model and that of MASTERS and GILBERT, 1983), since these two models have produced significantly different moment and depth values in other studies (e.g., ANDERSON and ZHANG, 1991; WALLACE *et al.*, 1991). The first step in the inversion procedure obtains a point source duration, τ, by minimizing a normalized error, σ; the second step determines the best depth and corresponding moment tensor by minimizing a normalized error, ρ, following ZHANG and KANAMORI (1988a). For the two attenuation models, Table 1 gives the inversion results for source duration, depth, seismic moment (M_0) and mechanism (strike ϕ, dip δ, rake λ) for the best double-couple of the moment tensor, the size of the non-double-couple component (Σ), and the total number of surface wave phases used in the inversion (N). Σ is the ratio of the smallest eigenvalue of the moment tensor in absolute value to the largest eigenvalue in absolute value, expressed as a percent (i.e., times 100). Thus, the percent of non-double-couple component is twice Σ.

The duration is unaffected by the different attenuation models but the depth and moment values are smaller for the Masters and Gilbert model. The range of values is much smaller than determined for the 1989 Macquarie Ridge earthquake (ANDERSON and ZHANG, 1991) probably because the difference between the attenuation models is enhanced for shallow events (ZHANG and KANAMORI, 1988a). For relatively deeper events such as the Kermadec earthquake, the Q effect on the depth estimate is small. The mechanism ($\phi = 275°$, $\delta = 60°$, $\lambda = 158°$) is quite stable and very similar to the Harvard solution (Figure 1).

Table 1

Boxcar time function

	Duration (s)	Depth (km)	Mechanism (ϕ, δ, λ)	Σ (%)	M_0 (10^{20} N-m)	N phases
Default Q	26	51	277/60/159	15	4.4	93
Masters & Gilbert Q	26	45	276/56/160	6	4.0	93
Default Q (No IDA stations)	20	45	276/60/157	7	4.4	71
Masters & Gilbert Q (No IDA stations)	20	40	275/56/158	6	4.2	71
Default Q (refined data)	26	44	276/64/157	7	4.2	74

The IDA data used in this study are relatively noisy so we performed the same inversions with 71 phases excluding the IDA data (Table 1). The effect of excluding the IDA data seems to be to reduce the duration to an even smaller value of 20 s and to reduce the depth by about 5 km. The mechanism remains stable and the moment values are slightly reduced. Although in general the size of the data set affects the solutions and their formal errors, we are unsure why the exclusion of the IDA data has these effects. It may be that this is just because the variance is slightly smaller when the IDA stations are excluded, or it may be related to the fact that the IDA stations have a longer-period response than the other stations.

Directivity

We investigated the spatial source finiteness by looking for rupture directivity adopting the method of ZHANG and KANAMORI (1988b) for both unilateral and bilateral rupture models. By minimizing the normalized error σ in the first-step inversion, we determined the best azimuths of a moving point source for seven periods from 150 to 300 s, assuming a fault length of 62.5 km and a rupture velocity of 2.5 km/s, which corresponds to a duration of 25 s. The variance reductions are small, and the rupture azimuths obtained from data of different periods span a large range for both unilateral (59°) and bilateral (44°) models, so no directivity was resolvable. This is not unexpected since the relatively short duration indicated by the simple point source analysis above is unlikely to produce major directivity effects.

We repeated these calculations excluding the IDA data for the reasons described above, and the results were similar. No directivity was resolved.

Results from Refined Data Set

At this stage we calculated the spectra and then carefully checked and deleted data which produced a much poorer fit (variance five times larger than the mean) to the spectral amplitude and phase plots at all periods. We also deleted R3 if both R1 and R2 were available for a station and the R3 error was larger than R1. This process reduced the number of phases to 74. We recalculated the source parameters using our simple boxcar time function and they differ only slightly from those determined with the larger data sets. Table 1 shows that in general the source parameters were stable with respect to the different attenuation models assumed, and the set of surface waves used. Figure 5 displays the normalized error, σ in the first-step inversion versus trial source durations at several periods, and the normalized error, ρ, in the second-step inversion versus trial source depths. Figure 6 shows the observed Rayleigh wave amplitude and phase spectra corrected for instrument and propagation effects, along with the spectra predicted from our derived source model at periods of 150, 256, and 300 s.

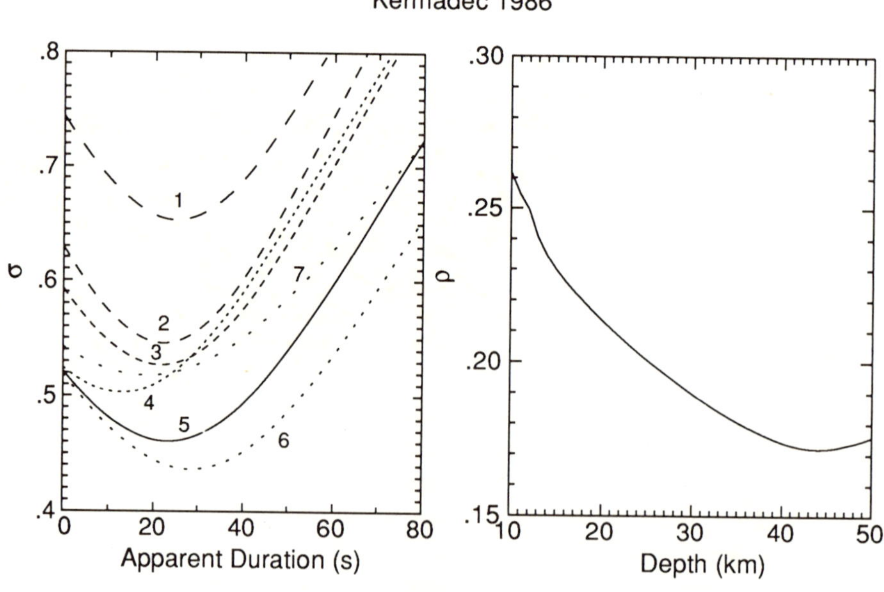

Figure 5

(a) The normalized error, σ, of the first-step inversion of surface waves, plotted versus duration (source process time) for the inversion with a simple source time function at the periods of (1) 150, (2) 175, (3) 200, (4) 225, (5) 256, (6) 275, (7) 300 s, respectively. The duration corresponding to the minimum error on each curve is taken as the value for that period. For the period 256 s (solid line) the minimum error corresponds to a source process time of 23 s. (b) The normalized error of the two-step inversion, ρ, plotted versus trial centroid depths for a simple source time function. The minimum error corresponds to a depth of 44 km.

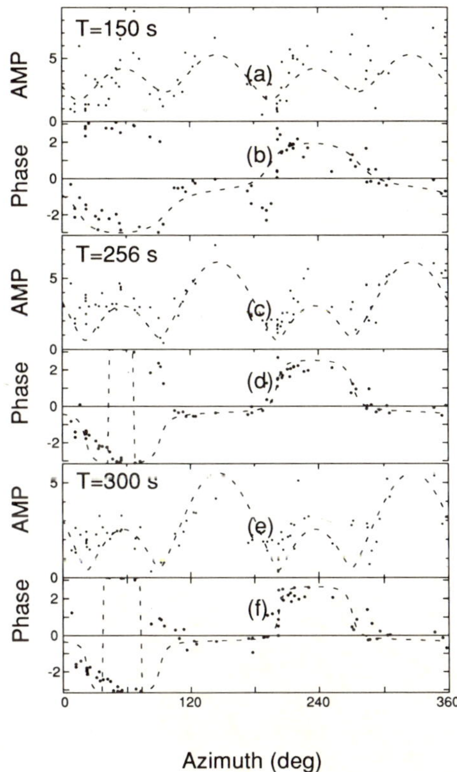

Figure 6

Observed propagation-corrected Rayleigh wave source spectra with periods of 150 s (a, b); 256 s (c, d); and 300 s (e, f). Amplitude spectra are given in units of 10^{25} m-s and phase spectra in radians. Dashed lines show source spectra predicted from the moment tensor solution obtained in this study.

Errors in the event location can propagate into errors in the resulting source parameters. We tested the effect of different epicentral locations by inverting the data, assuming locations up to 70 km to the north and 40 km to the south of the NEIC location. The centroid depth, mechanism, and non-double-couple component are relatively stable with respect to location. The largest effect is for the location 70 km to the north which yields a dip of the E-W trending nodal plane 13° greater than the NEIC location.

The non-double-couple component of the moment tensor is less than 15% of the total moment tensor for the last four solutions of Table 1. However, synthetic experiments show that the addition of a moment tensor similar to that of the mainshock to a moment tensor representing interplate underthrusting with similar moment would produce only a small non-double-couple component in the resulting moment tensor. Therefore the small non-double-couple component present in our moment tensor solution does not preclude the event having a complex mechanism with more than one focal mechanism.

The centroid time (13 s) is shorter than expected for an event of this size (M_W 7.7) from empirical relations between seismic moment and source duration (KANAMORI and ANDERSON, 1975; KANAMORI and GIVEN, 1981). LUNDGREN *et al.* (1989) also obtained a short duration (19 s) for this event from 6 Rayleigh wave spectra, as well as from their body wave deconvolution, and noted that the short duration fits the Rayleigh wave spectra better than longer durations. Other studies of similar magnitude events have shown that values of moment, centroid depth and duration derived from body wave inversions are often significantly less than those derived using surface waves. For the Kermadec earthquake, the source time function derived from body waves alone is sufficient to match the surface wave data and no significant long-period component is needed. This result is important because it indicates that a long source process time (duration) is not an inevitable result of the surface wave analysis.

Complex Time Function

In order to incorporate information about the source time function derived from the body waves into the surface wave inversion, we parameterized the moment rate function in terms of a series of isosceles triangles and a half-sine function. The approach follows ZHANG and LAY (1989a) and ANDERSON and ZHANG (1991). The general features of the time functions determined from the analysis of body waves, described below, can be modeled by 3 isosceles triangles with durations of 10 s starting at 0, 5 and 58 s and relative weights of 1.6, 1.0, and 0.9, respectively (see Figure 7a). The corresponding centroid time is 21 s. The first two triangles represent the first body-wave slip episode and have a centroid time of 7 s. A half-sine of duration τ is used to represent the long-period component. With this representation of the moment rate function a further parameter, γ, is necessary, giving the ratio of the half-sine component to the total moment rate function (e.g., for $\gamma = 0$ only the body-wave moment rate function, defined by the isosceles triangles, would contribute to the total moment rate function).

For each value of γ, the duration, τ, is found that minimizes σ, the error in the first-step inversion. Then γ and the best duration are used in the second-step inversion, which returns the depth and moment tensor that minimize the error, ρ. The results of this procedure are shown in Table 2 and Figure 7. In Figure 7b σ and ρ, the errors in the first and second steps of the inversion, respectively, are plotted versus γ. Figure 7c shows the results of the inversion-duration, τ, and moment, M_0, for different values of γ. Figure 7a shows the resulting time function for $\gamma = 0.5$ which was chosen somewhat arbitrarily for the smallest ρ that would not result in

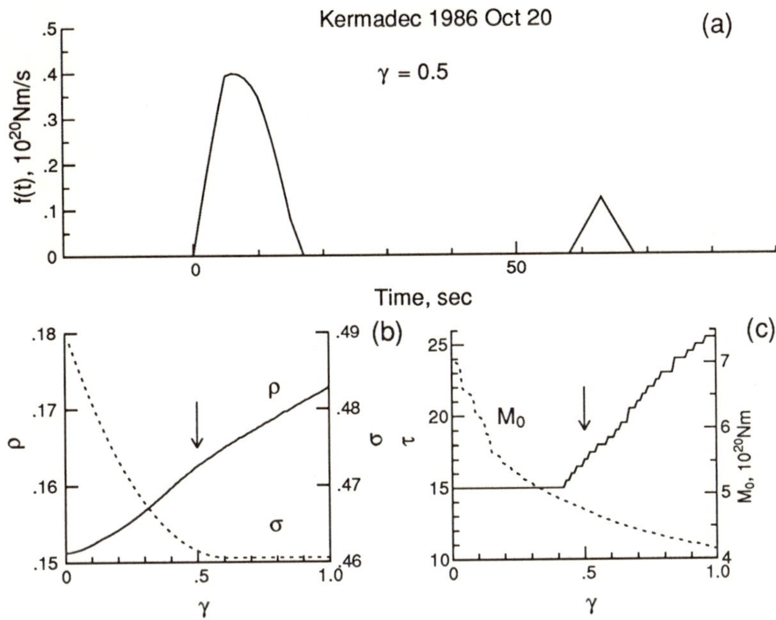

Figure 7

(a) Results of inversion of surface wave spectra using a moment rate function constructed by combining information inferred from the body-wave time function with a half-sine representing the long-period component. The onset time of the half-sine is fixed at 0 s. (b) The errors ρ (solid line) and σ (256 s period, dashed line) plotted against γ, the ratio of the long period half-sine component to the total moment. Arrows indicate the preferred value of γ chosen to keep ρ small while not accepting a significant increase in σ. (c) The duration of the half-sine τ (solid line) and total moment (dashed line) plotted against γ. The preferred half-sine has a duration of 17 s.

Table 2

Results of surface wave inversion with complex time function

	γ	Centroid of body wave time-function	τ duration (s)	Centroid time (s)	M_0 (10^{20} N-m)	Centroid depth (km)	Focal mechanism double-couple (ϕ, δ, λ)	Σ %
Two body wave slip episodes plus half-sine	0.5	21	17	15	4.7	38	275/61/156	10
Single body wave slip episode plus half-sine	0.2	7	83	14	4.6	40	275/62/156	4

γ is ratio of half-sine component to total moment rate function.

Σ is non-double-couple component.

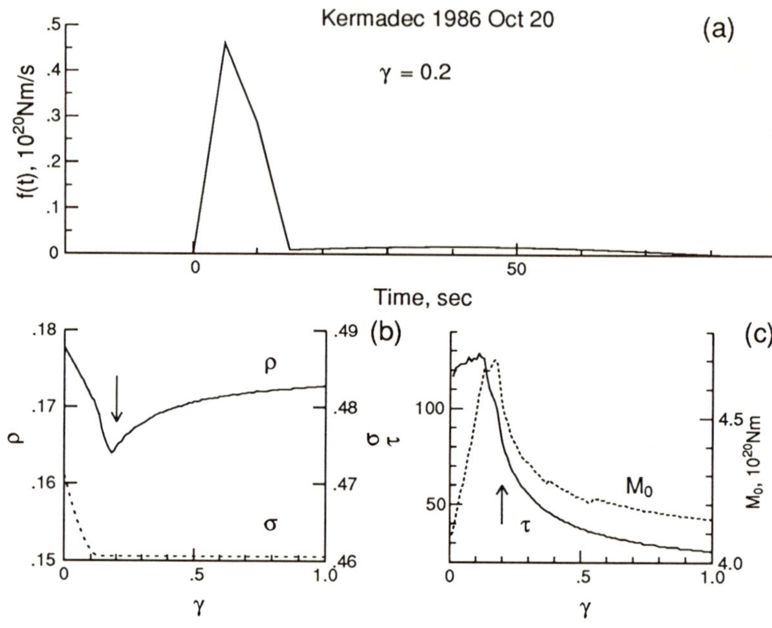

Figure 8

As in Figure 7 except that only the first 20 seconds of moment release in the time function obtained from body waves are included in the surface wave inversion. In this case the half-sine component has a duration of 83 s giving credence to the later episode of moment release determined from the body waves.

a significant increase in σ (Figure 7b). We fixed the minimum centroid time to be the same as the centroid time of the first body-wave moment episode because in previous trials the duration of the long-period component reduced to zero at small values of γ. This is expected because the centroid time of the body-wave time function (21 s) is longer than the centroid time (13 s) of the boxcar time function determined using surface waves. We interpret this result as indicating that no long-period component is required to explain the surface wave spectra.

We were unsure about the resolution of the second episode of moment release (at 58 s) so we repeated the inversions without it. The results are shown in Table 2 and Figure 8. It is noticeable that when no moment release is prescribed for the 58 s subevent then the duration of the half-sine is significant and extends slightly longer than the 70 s of the second body-wave episode in the model above. This supports the notion of a second moment release event which will be examined further in the next section.

Body Wave Analysis

The unusual long-period focal mechanism and the aftershock sequence, which include mainly underthrusting events typical for the plate boundary along the

Kermadec trench but only two events with focal mechanisms similar to the mainshock, raise the possibility that the mainshock consisted of complex faulting corresponding to more than one focal mechanism. In order to assess this possibility, we analyzed all available broadband body waveforms of the mainshock with an algorithm allowing multiple focal mechanisms. Additional goals were to estimate the spatial extent of rupture and to obtain a time function for use in the surface wave inversion. LUNDGREN *et al.* (1989) inverted 7 long-period *P* waves from GDSN, allowing only the rake to vary from the Harvard CMT solution, and obtained a solution very similar to the Harvard result and a depth of 40 km.

Data

We retrieved long-, intermediate-, and short-period records of the *P*, *SH*, and *PP* phases from the GDSN and NARS networks. We also included *P* and *SH* waves from KIP the one GEOSCOPE station that recorded a broadband teleseismic seismogram. We first constructed broadband displacements by simultaneously deconvolving the instrument responses and patching together the available short-, intermediate-, and long-period displacements in the frequency domain, then Fourier-transforming back to the time domain. Thereafter a 10–100 instrument response was convolved with the displacements, where the first number refers to the seismometer free period and the second number to the galvanometer free period. This instrument response was employed in order to (1) use an intermediate-period band to determine stable features of the rupture process, (2) emphasize frequencies near the corner frequency of the event, which have the most information about the rupture process, and (3) reduce very long- and short-period noise. Use of the 10–100 response is roughly similar to using velocity seismograms bandpass filtered between 10 and 100 seconds.

For the *P* waves, start times were picked from the short- and intermediate-period seismograms. It is often difficult to accurately pick the beginning of the *PP* or *S* phases. Start times of the *PP* phases were taken from the Jeffreys-Bullen travel times and further refined by comparison with *P* waveforms. They were, in some cases, quite uncertain and therefore were not included in final preferred inversions. We obtained accurate start times for the *SH* waves by comparing the *S-P* times for the large, impulsive aftershock of December 21, 1986 (for which the onsets of the *P* and *SH* phases were easy to pick) to the *S-P* times predicted by the iasp91 travel times. This technique assumes that the relative locations of the aftershock and mainshock are well-determined, which is probably the case. The actual *S-P* time for the aftershock was 5 seconds longer than predicted. Therefore, the start of the mainshock *SH* wave was estimated by adding 5 seconds to the *S-P* time predicted by iasp91 for the mainshock, given the differences in distance and depth from the relative relocations.

Structures near the source and receiver were approximated with flat layers including a 6 km-thick water layer above the source. Dipping bathymetry and structure can generate *P*-wave multiples as discussed by WIENS (1989). Although the effect of non-flat structures near the Kermadec source may be significant for stations in North America as argued by LUNDGREN *et al.* (1989), the actual structures near the trench are poorly known and may not be adequately approximated by dipping layers. The absolute location of the event is also poorly known, adding uncertainty to the estimation of the structure.

Variable Mechanism Inversions

To invert the body waveforms, we employed an iterative inversion technique developed by KIKUCHI and KANAMORI (1991) that returns the best-fitting focal mechanism for each iteration. This inversion fits the waveforms simultaneously by searching for the time, location, seismic moment, and focal mechanism of a point source that best correlates with the data, then subtracting the synthetics for such a point source from the data and repeating the procedure on the residual data. The method requires that the subevent time functions and fault plane grid be specified. Representative results for an inversion with a subevent time function consisting of a trapezoid with a rise time of 5 s and a duration of 8 s are shown in Figure 9. In this case the source was constrained to be a point source at 45 km (the best-fitting depth) to minimize the significant trade-offs that can occur between the various parameters. The focal mechanism returned by the first iteration for the slip that occurred from 0 to 15 s is roughly consistent with the first-motion solution and the long-period surface-wave solution described above. The second iteration often places a subevent at 58 s with a mechanism featuring thrusting on a north-south plane roughly consistent with a typical underthrusting mechanism or a compressional slab-bending event. However, the strike of this subevent is very unstable, rendering interpretation of the focal mechanism difficult. Its moment is about 45% of the moment of the first subevent. That a late episode of moment release occurred is evident from examination of the waveforms in Figure 9, as well from the displacement waveforms and from *SH* waveforms at stations NWAO and CTAO in Australia, which, unfortunately, saturated. A late episode of moment release is also suggested by the surface wave analysis discussed above.

More than 20 inversions were performed, varying the subevent time function, the allowed fault plane size, and the weighting of the data. Table 3 gives a list of the phases used. In general, the first iteration produced a fairly stable result roughly similar to that in Figure 1, although the mechanism could rotate around to approach a typical underthrusting mechanism. The subsequent iterations were less consistent. Reliably, either the second or third iteration would place a subevent about 60 s after the mainshock onset. In the inversions that allowed a range of depths for the point sources, this late subevent almost always located shallower

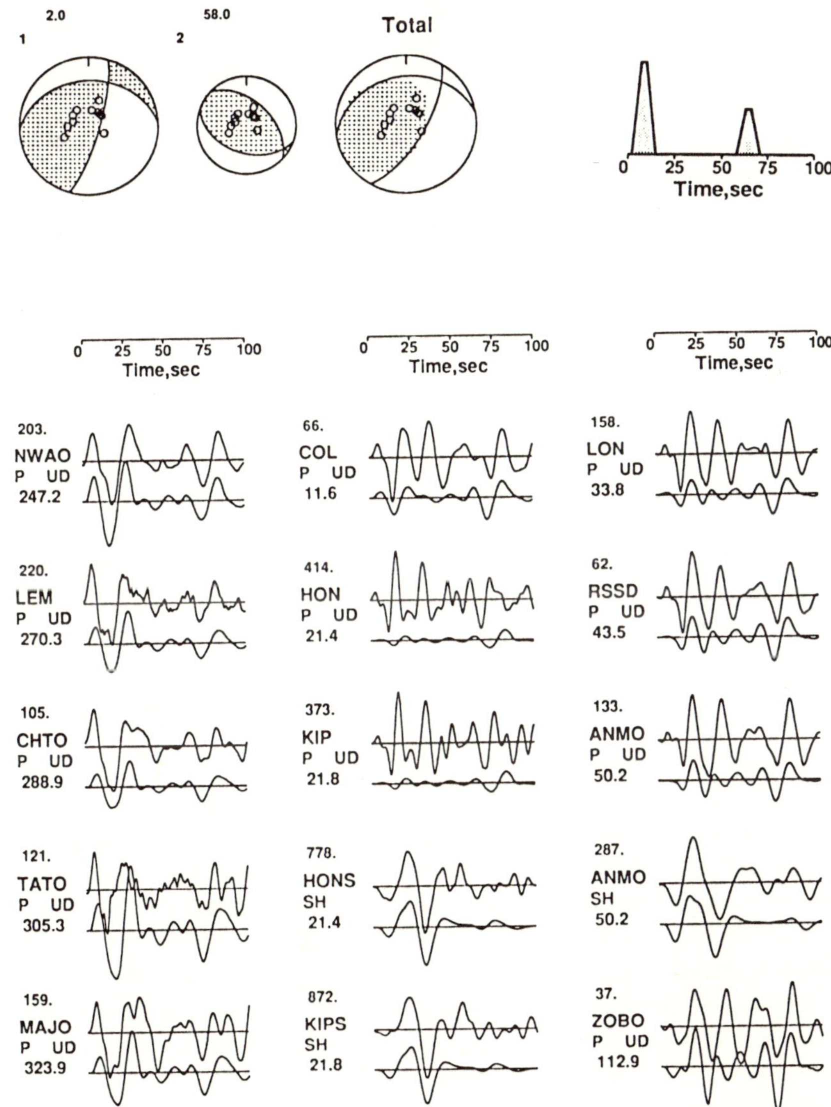

Figure 9

Results of inversion of teleseismic *P* and *SH* waves of the Kermadec earthquake for best-fitting focal mechanisms. The focal mechanisms and source time function after two iterations of the inversion; the time and iteration number are shown for each subevent. Comparison of synthetic waveforms (below) with data (above). Broadband displacements were convolved with a 10–100 instrument response with a gain of 1. Peak amplitude in microns (above) and azimuth (below) are given for each record.

Table 3

Station list for body waves

Station	Phase	Distance (°)	Azimuth (°)
NWAO	*P*	56.8	−112.8
LEM	*P*	74.7	−89.7
CHTO	*P*	94.4	−71.1
TATO	*P*	80.1	−54.7
MAJO	*P*	77.5	−36.1
COL	*P*	95.4	11.6
HON	*P & SH*	52.3	21.4
KIP	*P & SH*	52.5	21.8
LON	*P*	89.3	33.8
RSSD	*P*	97.5	43.5
ANMO	*P & SH*	90.9	50.2
ZOBO	*P*	97.5	112.9

than the first subevent, consistent with the notion that it resulted from slip on the shallow plate interface. The strike and dip of the late subevent mechanism were generally consistent with this possibility, but too variable to confirm it. At 15 to 30 s after the onset of the mainshock, however, a variety of mechanisms were found by the second, third, or fourth iterations in the various runs of the inversions, featuring large strike-slip components with nodal planes unrelated to those of the first-motion, CMT or surface wave results, or to the trench geometry. These subevents were difficult to interpret and may be more related to water multiples from a dipping sea bottom than to the earthquake source. Such water multiples would be expected to be stronger at stations in the eastern azimuths (e.g., North America) (LUNDGREN *et al.*, 1989), and the waveforms show large amplitudes at that time for the eastern stations. Considered together, the results of the different inversions indicate that the available distribution of body waves probably cannot resolve the fairly small difference between the long-period surface wave mechanism and a typical under-thrusting event along the Kermadec trench. However, the surface wave data are not compatible with an underthrusting mechanism overall.

Several weighting schemes for the records were explored. It was helpful to include the *SH* waves to constrain the mechanism, but their large amplitudes would have dominated the inversion had they not been down-weighted. Additionally, stations KIP and HON are located very close together and their records essentially duplicate each other. Furthermore, the *P* waves from KIP and HON contain anomalously short-period energy compared to the other *P* waves, suggesting that they were affected by some structure between Kermadec and Hawaii. Therefore, our preferred weighting scheme, used in Figure 9, was ANMO *SH*: 0.5, HON *P* and *SH*: 0.25, KIP *P* and *SH*: 0.25, and all other stations: 1.0. The inversion did prove somewhat sensitive to the weighting, diminishing our confidence in the robustness of the result. The addition of *PP* waves to improve the azimuthal coverage did not change the result significantly, and we felt that the improvement in azimuthal coverage was offset by the increased uncertainty in the start times of the *PP* waveforms.

In summary, multiple focal mechanisms during the first 20 s of moment release are not resolved, and the body waves appear to be adequately characterized by the long-period focal mechanism. The inversion consistently returned a different mechanism for the slip at 60 s, but a change of mechanism cannot be conclusively resolved then either. Considering the synthetic experiments on moment tensors mentioned in the surface wave section, the tectonically reasonable possibility that slip could have occurred simultaneously (i.e., within about 20 s) on the main plate interface and on an E-W trending plane cannot be definitively excluded. Such a situation would explain the unusual pattern of aftershocks. It is difficult to pursue this possibility further because the body waves are not sensitive enough to discriminate between the possible mechanisms, but there is no compelling evidence that this event was other than a simple double-couple with an unusual oblique mechanism.

Fixed Mechanism Inversions

To estimate the spatial extent of the rupture, we also performed a version of the iterative inversion (KIKUCHI and FUKAO, 1985) in which the focal mechanism is fixed. In some of these inversions with fixed mechanisms the *SH* waves were not included because their long-period character and uncertain start times are not helpful for locating slip. For similar reasons, *PP* waveforms were not included in the fixed mechanism inversions.

Several mechanisms were specified, including the body wave mechanism obtained in the inversion described above, the surface wave mechanism obtained above, the Harvard CMT mechanism, and the underthrusting mechanism of the largest aftershock (see Table 4). In the first three cases both the roughly N-S and E-W striking nodal planes were specified as the fault plane in different inversions because it is not obvious on which of the planes slip occurred, as neither is oriented in a typical geometry for subduction zones. The spatial extent of rupture was fairly limited for all these cases, and most of the slip was concentrated in a relatively small area, about 40 by 40 km, near the hypocenter. When the roughly E-W striking fault planes were specified, the region of maximum moment release was centered about 20 km west of the hypocenter. The residual errors (shown in Table 4) were smaller for E-W striking nodal planes than for the N-S trending planes, supporting our choice of the E-W plane as the fault plane. It is interesting to note that a representative underthrusting mechanism explains the data almost as well as the best oblique mechanisms. As mentioned above, the body waves appear unable to resolve this difference, but an oblique mechanism is required by the surface wave data. Time functions similar to that shown in Figure 7 were obtained in all cases.

Table 4

Fixed mechanism inversions‡

	Strike	Dip	Rake	Error*	Error†
Our surface wave	275	61	156	0.4350	0.3987
mechanism	17	69	31	0.4533	0.4115
Our body wave	268	33	154	0.3509	0.4220
mechanism	20	76	60	0.3995	0.4357
Harvard CMT	270	56	158	0.3881	0.3985
mechanism	13	71	36	0.4200	0.4214
Underthrusting	207	23	106	0.3999	0.4125
mechanism					

‡ 8 iterations of the inversion using a trapezoid with 4 s rise time and 7 s duration and an extended fault plane 160 km along strike by 50 km along dip
* error obtained including *SH* waves
† error obtained excluding *SH* waves

Stress Drop and Source Spectrum

We calculated the static stress drop for the Kermadec earthquake, assuming a circular rupture area. The static stress drop of a circular crack is given by:

$$\Delta\sigma = \frac{7\pi\mu\bar{D}}{16a} = \frac{7M_0}{16a^3},$$

where $\bar{D}$ is the average displacement over surface area $S = \pi a^2$ and M_0 is the moment (KANAMORI and ANDERSON, 1975). Using $a = 20$ to 25 km and $M_0 = 4 \times 10^{20}$ N-m (both estimated from the fixed mechanism inversions) yields a static stress drop of 100 to 200 bars, which is five to ten times larger than is typical for interplate underthrusting earthquakes.

We computed an average source spectrum (Figure 10) from the teleseismic *P* waves and derived from it estimates of the seismically-radiated energy and the Orowan stress drop following HOUSTON (1990). The energy E is proportional to the area under the squared velocity spectrum, and the Orowan stress drop (twice the apparent stress) is given by:

$$\Delta\sigma_0 = \frac{2\mu E}{M_0}.$$

We calculate an energy of 1×10^{23} ergs and obtain an Orowan stress drop of 30 bars. This is high compared to similarly-determined stress drops of interplate subduction earthquakes (HOUSTON, 1990), as is the static stress drop, consistent with intraplate faulting with a relatively long repeat time rather than interplate faulting, which typically has a much shorter repeat time.

1986 Kermadec Spectrum

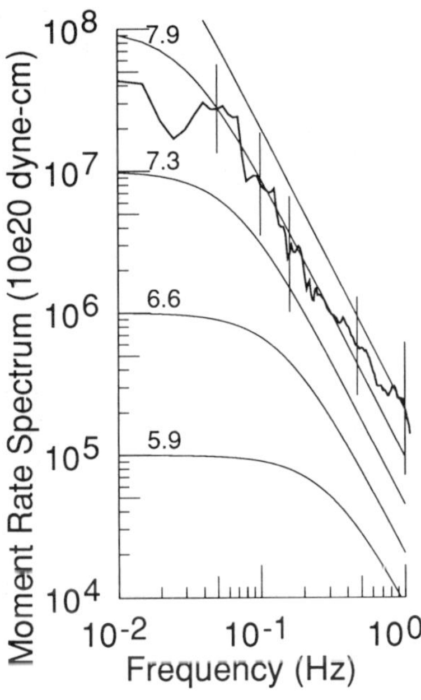

Figure 10

Source spectrum of the Kermadec earthquake from P waves. Thin solid lines show the ω^{-2} model with a stress drop of 30 bars for reference. The stress drop of this event is 30 bars. The long-period spectral amplitude is fixed at the seismic moment.

The Largest Aftershock

The largest aftershock ($M_W = 6.2$) occurred two months after the mainshock on December 21, 1986. Its Harvard CMT solution is that of a typical underthrusting mechanism and the CMT depth is 17 km. The waveforms have a very simple, impulsive character. To check the depth, we inverted 11 broadband teleseismic P waves for the time function of a point source with the CMT mechanism at different depths. The lowest residual occurred at a depth of 17 km, confirming the CMT depth, and emphasizing the difference between this aftershock (and by implication the other shallow, underthrusting aftershocks) and the mainshock, which was significantly deeper.

Discussion

A Brief, High Stress Drop Intraplate Event

Generally, the source parameters of the Kermadec earthquake derived using the surface and body wave techniques are similar. The centroid depth is 45 ± 5 km, the moment is $4.5 \pm 0.3 \times 10^{20}$ N-m, and the mechanism is oblique-compressional slip (Figure 1). The depth, focal mechanism, and relation to aftershock locations of this event suggest intraplate faulting without a mechanism change, although the part of the body waveforms from 15 to 40 s remain poorly explained. Perhaps this is partly or entirely due to water multiples; however, the possibility of some rupture complexity cannot be excluded. The short duration and high stress drop are also indicative of intraplate faulting. As a relatively shallow oblique-slip subduction zone event, the Kermadec earthquake is unique for its size (LUNDGREN et al., 1989).

The centroid time of 13 s indicates an anomalously brief duration (by about a factor of two) of moment release for an event of this size. The method employed for the surface wave analysis in this study has been applied to several other large earthquakes. Generally, a significant long-period component of rupture is needed to explain the surface wave observations, with source process times (e.g., the duration of the half-sine component) of 93, 140, and 165 s for the 1989 Macquarie, 1977 Sumba, and the nearby 1977 Tonga earthquakes (ANDERSON and ZHANG, 1991; ZHANG and LAY, 1989b). Our determination of a half-sine duration of 17 s for the Kermadec earthquake, however, rules out any significant slow slip during its rupture. Thus, the analysis of this earthquake shows that the method does not invariably return a long source process time, which suggests that the long-period slip often obtained from surface wave studies is meaningful.

Figure 11 and Table 5 show duration (twice centroid time) and seismic moment computed from surface wave spectra for 14 events, using boxcar time functions

Table 5

Moment and duration for large shallow earthquakes

Event	Moment (N-m)	Duration (sec)
6/22/77 Tonga	1.70×10^{21}	84
8/19/77 Sumba	5.60×10^{21}	80
12/12/79 Colombia	2.10×10^{21}	118
7/17/80 Santa Cruz Is.	6.80×10^{20}	88
9/01/81 Samoa	5.40×10^{20}	44
6/19/82 El Salvador	1.40×10^{20}	34
5/26/83 Akita-Oki	8.00×10^{20}	46
11/30/83 Chagos	8.30×10^{20}	62
3/03/85 Chile	1.28×10^{21}	70
9/19/85 Michoacan	1.05×10^{21}	76
10/20/86 Kermadec	4.20×10^{20}	26
5/23/89 Macquarie	2.15×10^{21}	46
10/18/89 Loma Prieta	3.40×10^{19}	11
5/20/90 Sudan	7.00×10^{19}	20

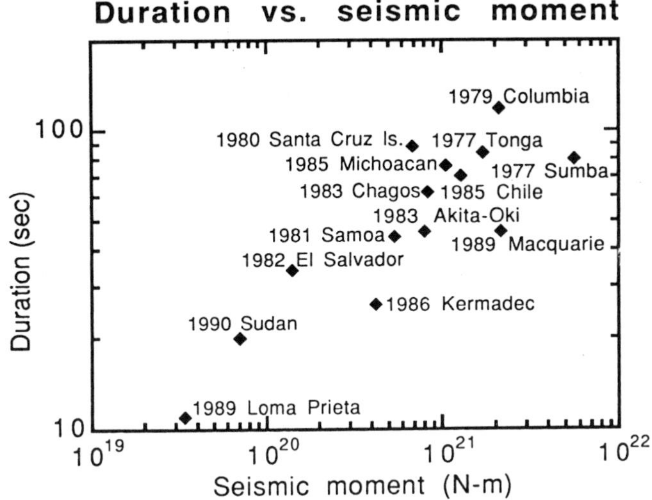

Figure 11

Duration vs. seismic moment for 14 large earthquakes, determined from surface wave analyses. The duration is twice the centroid time for a boxcar time function. The plot shows the general increase in duration with the cube root of moment, as well as the anomalously brief duration of the Kermadec earthquake for its size.

(e.g., ZHANG and KANAMORI, 1988a,b; ZHANG and LAY, 1989a,b; ANDERSON and ZHANG, 1991; VELASCO *et al.*, 1993). Four of these events can be considered anomalous—the 1989 Loma Prieta, 1986 Kermadec, 1989 Macquarie, and 1977 Sumba earthquakes. Recall the empirical relation that the seismic moment is proportional to the cube of the source duration τ (KANAMORI and ANDERSON, 1975). For the Kermadec earthquake we obtained a ratio $M_0/\tau^3 = 2.4 \times 10^{16}$ N-m/s^3, which is similar to the ratio of 2.3×10^{16} N-m/s^3 for the 1989 Loma Prieta earthquake ($M_0 = 3.0 \times 10^{19}$ N-m, $\tau = 11$ s (VELASCO *et al.*, 1993)). However, for typical low-angle thrust earthquakes this ratio was found to be 2.5×10^{15} N-m/s^3 (FURUMOTO and NAKANISHI, 1983). This is consistent with Figure 11 and implies that the Kermadec event is a fast earthquake with a rupture time about half as long as that for subduction zone earthquakes with similar seismic moments.

Tectonic Significance of this Event

It is difficult to determine which of the two nodal planes (N-S or E-W) is the actual fault. The lack of observable directivity associated with the surface waves indicates a small source region and does not allow us to uniquely determine the fault plane. The body-wave inversions weakly favor the E-W plane but do not provide overwhelming support for this interpretation. The N-S elongation of the aftershocks would seem to favor the N-S plane (preferred by LUNDGREN *et al.*, 1989), but, as discussed previously and below, most of the aftershocks appear to be triggered slip on the underthrusting plane and not aftershocks that define the fault plane. We favor the E-W plane as the fault plane and suggest that this event represents a lateral segmentation or tear in the subducting plate.

Several factors suggest that slip in the mainshock occurred along the E-W trending nodal plane, rather than along the roughly N-S plane. First, there is a right-lateral offset in the free-air gravity anomaly associated with the volcanic arc that would correspond to the right-lateral direction of slip associated with the E-W nodal plane (Figure 12). The arc is offset perpendicular to the trench near the mainshock epicenter. Although the offset is in the upper plate, it probably reflects the configuration of the subducting plate. Also, the strike of the trench changes near the epicenter: north of 28.36°S the trench strikes N10°E compared to N20°E south of that latitude. Furthermore, the relocated aftershocks, although quite diffuse, are generally shallower in the northern part of the aftershock zone than in the southern part (Figure 4). These factors suggest that the subducting lithosphere may be segmented near the site of the Kermadec earthquake. Such lateral segmentation could be related to a buoyant Louisville Ridge, which intersects the trench about 200 km to the north and may act as an obstacle to subduction.

Another factor suggesting that the E-W nodal plane is the fault plane is the distribution of events with mechanisms similar to that of the mainshock. A search of all CMT solutions in the region from January 1, 1977 to June 30, 1991 yielded only three events with focal mechanisms similar to that of the mainshock (see Figure 3). Two of these occurred before the mainshock; one is an aftershock. In contrast to the entire set of aftershocks which extend about 150 km north and 50 km south of the mainshock, the three events with focal mechanisms similar to the mainshock are all located within about 20 km of the mainshock epicenter. Furthermore, all three events are fairly deep, with Harvard CMT depths of 57 km (July 24, 1983), 43 km (July 28, 1983), and 63 km (December 20, 1986), in contrast to the underthrusting aftershocks, which are mostly shallower than 40 km. This is suggestive of an E-W striking structure within the subducting plate similar in location, depth, and strike to the E-W striking nodal plane of the mainshock. This also suggests that the underthrusting aftershocks are not true aftershocks, but represent slip on the interplate boundary triggered by strain transferred to the adjoining trench segments. Perhaps these aftershocks extended preferentially to the north of the mainshock because the stress on the interplate boundary south of the mainshock had already been released by the large, presumably underthrusting events in 1976. It is also possible that these underthrusting aftershocks are true aftershocks of a relatively small and unresolved underthrusting event "hidden" in the mainshock. It appears that the N-S distribution of the aftershocks is a result of their alignment parallel to the trench and is not therefore a convincing indication of mainshock faulting on the N-S striking plane.

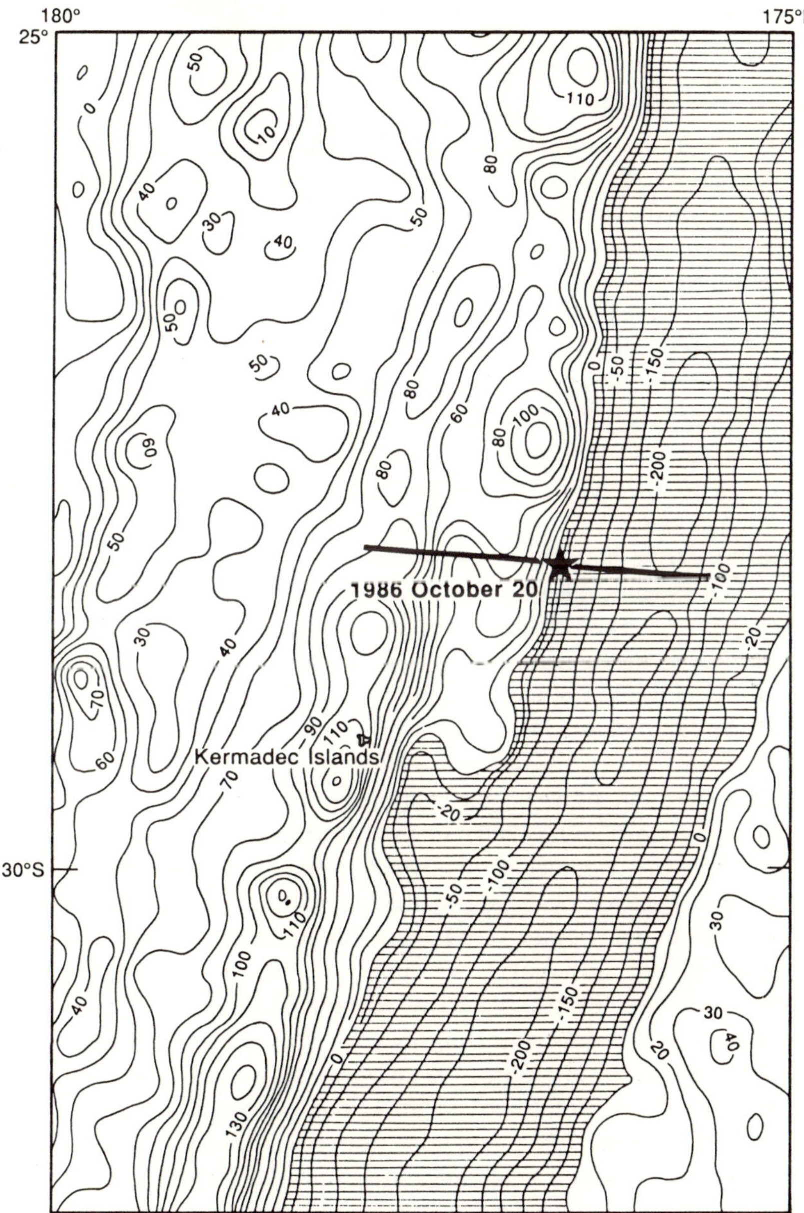

Figure 12

Free-air gravity anomalies (in mgals) in the region of the Kermadec trench from DAVY (1989). The star shows the epicenter of the 1986 Kermadec earthquake; the heavy line shows the strike of the inferred fault plane. Note the change in strike of the trench (negative gravity anomaly) near the epicenter of the Kermadec earthquake, and the offset of the positive gravity anomaly associated with the volcanic arc.

In our interpretation, the mainshock occurred in the Pacific plate on the E-W striking plane as a slab tear earthquake. While the paucity of aftershocks with mechanisms similar to the mainshock and on the mainshock fault plane is unusual, it is generally thought that intraplate events tend to have fewer true aftershocks than interplate events of comparable size. Similar triggering of shallow underthrusting earthquakes on the plate interface was observed following the 70-km-deep normal-faulting earthquake of June 22, 1977 in Tonga (CHRISTENSEN and LAY, 1988).

We note that the horizontal projection of the slip vector for the E-W striking plane is very close to the relative plate motion direction predicted by the NUVEL-1 pole (272°, DEMETS *et al.*, 1990). Because the event occurred directly beneath the coupled plate interface, perhaps the slip vector of this intraplate event was controlled by the plate motions. Since the downgoing plate is not yet decoupled from the overriding plate in this region, the concordance of P or T axis orientation with slab dip that is often recognized in deeper slab events would not be expected, and does not occur (Figure 1).

Slab tearing events have occurred further south along the Kermadec trench in New Zealand (REYNERS, 1991), and segments of the Benioff zone with significantly different geometry, marked by sharp vertical discontinuities, have also been recognized in the New Zealand region (ROBINSON, 1986). Although large earthquakes associated with the process of lateral segmentation of subducting plates are rare, we believe that the 1986 Kermadec earthquake is most easily interpreted as a significant slab tearing event.

From another point of view, the Kermadec earthquake can be considered a stress meter in the downgoing plate. In this interpretation the orientations of the P and T axes are more significant than which nodal plane is identified as the fault plane. The tectonic setting of the Kermadec earthquake, an intraplate event in the downgoing slab on the seaward edge of the seismogenic interface, is similar to that of an outer-rise earthquake. We note that the T axis of the Kermadec earthquake is pointing toward the 1976 earthquake doublet (Figures 2 and 3), which presumably were underthrusting events. A crude analogy may be drawn with the 1987–88 Gulf of Alaska earthquake sequence. LAHR *et al.* (1988) observed that the T axes of three shallow strike-slip earthquakes in the Pacific plate ($M_S = 6.9$, 7.6, and 7.6) point northwest towards the rupture zone of the great Alaskan earthquake of 1964, suggesting that the Pacific plate is accommodating tensional stresses induced by that event. The P axes of the Gulf of Alaska events may be controlled by the oblique collision and subduction of the buoyant Yakutat terrane. If this interpretation can be applied to the Kermadec earthquake, then the orientation of its T axis suggests that the event accommodated tensional stresses induced by the 1976 events, while the orientation of its P axis towards the northwest could suggest that the interplate boundary north of the epicenter is locked.

The Role of the Louisville Ridge

The morphology of the subducting plate in the Tonga-Kermadec zone is very complex at depth (BILLINGTON, 1980; FISCHER *et al.*, 1991) and several possible segment boundaries have been proposed (BURBACH and FROHLICH, 1986; PELLETIER and LOUAT, 1989). We, therefore, examined the stress state of the entire subducting slab in the region of the Kermadec earthquake using focal mechanisms from the Harvard CMT catalog. Figure 13 shows the projections onto a horizontal plane of compression and tension axes of events at all depths in this region during the period 1977 to 1991. A change in the nature of the stress field is evident near the epicenter of the Kermadec earthquake at about latitude 28°S. This change is most noticeable in the projections of the T axes, which become shorter (i.e., T axes are more vertical) south of that latitude. In Figure 14 cross sections of P axes north and south of 28°C are projected onto a vertical plane perpendicular to the trench, and reveal a distinct segmentation in the subducted slab just south of the 1986 Kermadec earthquake. There is a clear change in the dip of the seismicity and the P axes between the northern and southern cross section. Although the change is more clearly manifested below 300 km depth, the 1986 Kermadec earthquake appears to be part of the process of this segmentation. The connection of shallow processes to the morphology of the slab at greater depths is also seen in the Santa Cruz Islands subduction zone, where segmentation inferred from the intermediate-

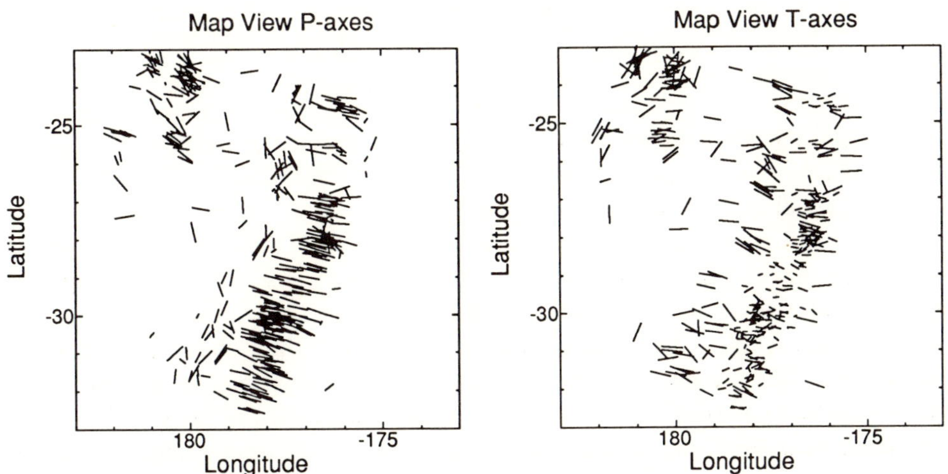

Figure 13

Map views of stress axes of all CMT catalog events in the region of the Kermadec earthquake. The P (left) and T (right) axes are projected onto the horizontal. Note the change in the character of the stress field near the latitude of the Kermadec earthquake.

Cross-sections of stress axes

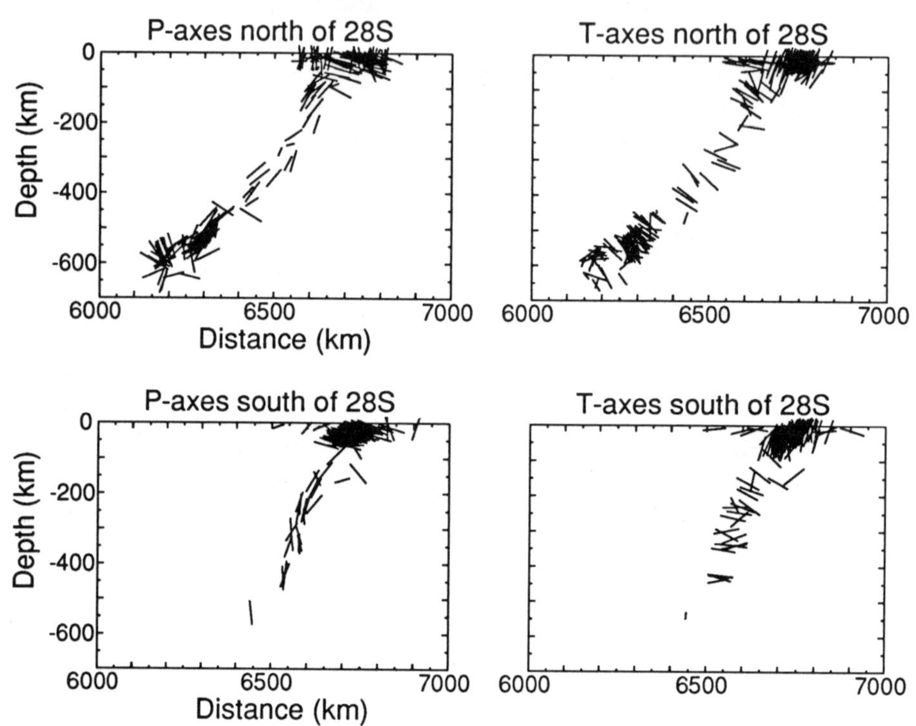

Figure 14

Cross sections of the *P* axes of the events in Figure 13 north and south of latitude 28°S near the Kermadec earthquake. The axes are projected onto a vertical plane perpendicular to an arc approximating the trench (pole at latitude 0°, longitude 127°E).

depth seismicity seems to control the mechanical condition of the plate interface at shallow depth (TAJIMA *et al.*, 1990).

Geophysical and geochemical observations suggest that the segmentation seen in Figures 13 and 14 is directly related to the subduction of the Louisville Ridge. At, and to the north of, its intersection with the trench, the Louisville Ridge has strongly affected trench morphology (BALLANCE *et al.*, 1989; McCANN and HABERMANN, 1989). This is clearly seen at 26°S where the Osbourne seamount is impinging on the trench (Figure 2). 200 km north of this intersection the trench achieves a depth of 10,800 m, the second deepest bathymetry in the world, while at the intersection itself the trench shallows to 6,000 m. The timing and geometry of the back-arc opening in the Lau Basin suggest a relationship between rapid extension in the basin and southward progression of the intersection of the trench with the Louisville Ridge. Based on the petrogenesis and comparative chemistry of lava samples dredged from the Lau Basin, VALLIER *et al.* (1991) conclude that Lau

Basin opening has moved progressively southward, and note that the opening of the Lau Basin is coincident with passage of the Louisville Ridge at the trench.

The geometry and bathymetry of the trench near its intersection with the Louisville Ridge (Figure 2) suggests that the Louisville Ridge is buoyant and impedes subduction locally. Once subducted to a depth of about 50 km, the ridge material is expected to transform to eclogite, losing its anomalous buoyancy (AHRENS and SCHUBERT, 1974). Subduction of the adjacent plate north and south of the ridge material appears not to be significantly impeded. This situation is likely to produce stresses that tend to segment the subducting plate along strike. SCHOLL et al. (1992) suggest that subduction of the Louisville Ridge conditions the downgoing plate by displacing the thermal regime so that slab rollback occurs more readily and rapidly. This enhanced rollback, which would produce seaward migration of the trench with respect to the mantle, would then allow the progressive opening of the Lau Basin, and would also explain the shallower dip of the Tonga slab compared to the Kermadec slab. North of the intersection the slab is contorted at depth, while to the south it is simple and more steeply dipping.

The Louisville Ridge appears to have a broad impact zone. It is not simply a chain of discrete volcanos, but a broad topographic swell with crustal height of 3 to 4 km (LONSDALE, 1986; WATTS et al., 1986). From a bathymetric profile along the Tonga-Kermadec trench showing the width of the hot-spot generated swell associated with the Louisville Ridge (Figure 12 of TAGUDIN et al., 1992), one can infer a width of impact on the trench of over 400 km. It is, therefore, quite plausible that the Louisville Ridge affects the process of subduction near the epicenter of the Kermadec earthquake 200 km south of the point where it impinges directly on the trench at 26°S.

Another possibility, which we consider less likely, is that the segmentation could be related to the Eltanin fracture zone, a less prominent feature on the subducting plate, which appears to intersect the trench near 29°S (HERZER et al., 1987). In either case, Figures 13 and 14 suggest that the Kermadec earthquake can be considered part of the process of plate segmentation.

Conclusions

The focal mechanisms determined from long-period surface waves, from the broadband body waves, and from first motions agree roughly and are consistent with oblique faulting within the plate that does not fit easily into the regional tectonic framework. The depth, short duration, relatively high stress drop, and paucity of true aftershocks are also indicative of intraplate faulting. The fact that the few events with mechanisms similar to the mainshock all occurred near the mainshock epicenter, a change in trench strike near the latitude of the mainshock epicenter, and inversions of teleseismic body waveforms, favor the E-W plane for the fault plane. Despite some complexity in the body waves, multiple focal mechanisms are not required or resolved. An oblique-compressional event of this size within the shallow downgoing lithosphere is rare in a subduction-zone setting. The Kermadec earthquake appears to be associated with segmentation of the subducting plate produced by forces related to the subduction of the Louisville Ridge.

Acknowledgments

We thank L. Ruff, M. Kikuchi, and R. Benites for reviews. L. Ruff suggested the interpretation of the Kermadec earthquake as a stress meter. This research was supported by N.S.F. Grant EAR-8721190 to H.H., as well as by the Institute of Geophysics and Planetary Physics, at Lawrence Livermore National Laboratory. H.A. was supported by the United States–New Zealand Co-operative Science Program and the Fulbright-Hayes Program. Facility support was provided by a grant from the W. M. Keck Foundation. Contribution #188 of the Institute of Tectonics and C. F. Richter Seismological Laboratory, UCSC, and #11 of the Institute of Geological and Nuclear Sciences, New Zealand.

REFERENCES

AHRENS, T., and SCHUBERT, G. (1975), *Gabbro-eclogite Reaction Rate and its Geophysical Significance*, Rev. Geophys. Space Phys. *13*, 383–400.

ANDERSON, H., and ZHANG, J. (1991), *Long-period Seismic Radiation from the May 23, 1989, Macquarie Ridge Earthquake: Evidence for Coseismic Slip in the Mantle?* J. Geophys. Res. *96*, 19853–19863.

BALLANCE, P. F., SCHOLL, D. W., VALLIER, T. L., and HERZER, R. H. (1989), *Subduction of a Late Cretaceous Seamount of the Louisville Ridge at the Tonga Trench: A Model of Normal and Accelerated Tectonic Erosion*, Tectonics *8*, 953–962.

BILLINGTON, S. (1980), *The Morphology and Tectonics of the Subducted Lithosphere in the Tonga-Fiji-Kermadec Region from Seismicity and Focal Mechanism Solutions*, Ph.D. Dissertation, Cornell Univ., Ithaca, N.Y.

BURBACH, G. v. N., and FROHLICH, C. (1986), *Intermediate and Deep Seismicity and Lateral Structure of Subducted Lithosphere in the Circum-Pacific Region*, Rev. Geophys. *24*, 833–874.

CHRISTENSEN, D., BECK, S., and GALEWSKY, J. (1991), *The January 14, 1976 Kermadec Earthquake Sequence* (abstract), EOS Trans. Am. Geophys. Union, 348.

CHRISTENSEN, D. H., and LAY, T. (1988), *Large Earthquakes in the Tonga Region Associated with Subduction of the Louisville Ridge*, J. Geophys. Res. *93*, 13367–13389.

DAVY, B. W., *Southern Oceans Sheet 1; Gravity Anomaly Map*, Miscellaneous Series, 1:4000000, Wellington, New Zealand, Department of Scientific and Industrial Research, 1989.

DEMETS, C., GORDON, R. G., ARGUS, D. F., and STEIN, S. (1990), *Current Plate Motions*, Geophys. J. Int. *101*, 425–478.

DEWEY, J. W. (1972), *Seismicity and Tectonics of Western Venezuela*, Bull. Seismol. Soc. Am. *62*, 1711–1751.

DOUTCH, H. F. (1981), *Plate Tectonic Map of the Circum-Pacific Region, Southwest Quadrant*, American Assoc. Petroleum Geologists, Tulsa, U.S.A.

DZIEWONSKI, A. M., and ANDERSON, D. L. (1981), *Preliminary Reference Earth Model*, Phys. Earth Planet. Inter. *25*, 297–356.

DZIEWONSKI, A. M., and STEIM, J. M. (1982), *Dispersion and Attenuation of Mantle Waves through Waveform Inversion*, Geophys. J. R. Astr. Soc. *70*, 503–527.

FISCHER, K. M., CREAGER, K. C., and JORDAN, T. H. (1991), *Mapping the Tonga Slab*, J. Geophys. Res. *96*, 14403–14427.

FURUMOTO, M., and NAKANISHI, I. (1983), *Source Times and Scaling Relations of Large Earthquakes*, J. Geophys. Res. *88*, 2191–2198.

HERZER, R. H., BALLANCE, P. F., SCHOLL, D. W., and PONTOISE, B. (1987), *Subduction of an Aseismic Ridge in Sediment-starved Trench—Effects on the Ridge, the Trench Wall, and the Forearc* (abstract), Geol. Soc. N.Z. Misc. Publ. *37A*, 56.

HOUSTON, H. (1990), *Broadband Source Spectrum, Seismic Energy, and Stress Drop of the 1989 Macquarie Ridge Earthquake*, Geophys. Res. Lett. *17*, 1021–1024.

HOUSTON, H., and BECK, S. (1990), *A Broadband Waveform Analysis of the Rupture Process of the October 20, 1986 Kermadec Earthquake ($M_W = 7.7$)*, Seismol. Res. Lett. *61*, 27.

HOUSTON, H., ANDERSON, H., and ZHANG, J. (1990), *The 20 October 1986 Kermadec Earthquake: An Anomalous Event*, EOS Trans. Am. Geophys. Union *71*, 1468.

KANAMORI, H., and GIVEN, J. W. (1981), *Use of Long-period Surface Waves for Rapid Determination of Earthquake-source Parameters*, Phys. Earth Planet. Inter. *27*, 8–31.

KIKUCHI, M., and FUKAO, Y. (1985), *Iterative Deconvolution of Complex Body Waves from Great Earthquakes—the Tokachi-Oki Earthquake of 1968*, Phys. Earth Planet Inter. *37*, 235–248.

KIKUCHI, M., and KANAMORI, H. (1991), *Inversion of Complex Body Waves—III*, Bull. Seismol. Soc. Am. *81*, 2335–2350.

LAHR, J. C., PAGE, R. A., STEPHENS, C. D., and CHRISTENSEN, D. H. (1988), *Unusual Earthquakes in the Gulf of Alaska and Fragmentation of the Pacific Plate*, Geophys. Res. Lett. *15*, 1483–1486.

LONSDALE, P. (1988), *Geography and History of the Louisville Hotspot Chain in the Southwest Pacific*, J. Geophys. Res. *93*, 3078–3104.

LUNDGREN, P. R., OKAL, E. A., and WIENS, D. A. (1989), *Rupture Characteristics of the 1982 Tonga and 1986 Kermadec Earthquakes*, J. Geophys. Res. *94*, 15521–15539.

MCCANN, W. R., and HABERMANN, R. E. (1989), *Morphologic and Geologic Effects of the Subduction of Bathymetric Highs*, Pure and Appl. Geophys. *129*, 41–69.

MASTERS, G., and GILBERT, F. (1983), *Attenuation in the Earth at Low Frequency*, Philos. Trans. R. Soc. London, A *308*, 479–522.

PELLETIER, B., and LOUAT, R. (1989), *Seismotectonics and Present-day Relative Plate Motions in the Tonga-Kermadec-Havre Region*, Tectonophys. *165*, 237–250.

REGAN, J., and ANDERSON, D. L. (1984), *Anisotropic Models of the Upper Mantle*, Phys. Earth Planet. Inter. *35*, 227–263.

REYNERS, M., GLEDHILL, K., and WATERS, D. (1991), *Tearing of the Subducted Australian Plate During the Te Anau, New Zealand, Earthquake of 1988 June 3*, Geophys. J. Int. *104*, 105–115.

ROBINSON, R. (1986), *Seismicity, Structure and Tectonics of the Wellington Region, New Zealand*, Geophys. J. R. Astr. Soc. *87*, 379–409.

ROMANOWICZ, B. A., and GUILLEMANT, P. (1984), *An Experiment in the Retrieval of Depth and Source Mechanism of Large Earthquakes using Very Long-period Rayleigh Wave Data*, Bull. Seismol. Soc. Am. *74*, 417–437.

SCHOLL, D. W., TAGUDIN, J. E., and VALLIER, T. L. (1992), *Extensional and Thermal Consequences of Subduction of the Louisville Ridge at the Tonga Trench and the Basinward Migration of the Arc Magmatic Front and Along-arc Progression of Arc-sector Dormancy* (Abstract), Volcanism Associated with Extension at Consuming Plate Margins. British Antarctic Survey, Cambridge, England.

SYKES, L. R., ISACKS, B., and OLIVER, J. (1969), *Spatial Distribution of Deep and Shallow Earthquakes of Shallow Magnitude in the Fiji-Tonga Region*, Bull. Seismol. Soc. Am. *59*, 1093–1113.

TAGUDIN, J. E., and SCHOLL, D. W., *The westward migration of the Tofua Arc toward Lau Basin*. In *Contributions to the Marine and On-land Geology and Resource Assessment of the Tonga-Lau-Fiji Region* (Ballance, P. F., Herzer, R. H., and Stevenson, A. J. eds.) (Technical Series, CCOP, Suva, Fiji 1992).

TAJIMA, J., RUFF, L. J., KANAMORI, H., ZHANG, J., and MOGI, K. (1990), *Earthquake Source Processes and Subduction Regime in the Santa Cruz Islands Region*, Phys. Earth Planet. Inter. *61*, 269–290.

VALLIER, T. L., and 15 others (1991), *Subalkaline Andesite from Valu Fa Ridge, a Back-arc Spreading Center in Southern Lau Basin: Petrogenesis, Comparative Chemistry, and Tectonic Implications*, Chemical Geology *91*, 227–256.

VELASCO, A. A., LAY, T., and ZHANG, J. (1993), *Long-period Surface Wave Inversion for Source Parameters of the 18 October 1989, Loma Prieta Earthquake*, Phys. Earth Planet. Inter. *76*, 43–66.

WALLACE, T. C., VELASCO, A., ZHANG, J., and LAY, T. (1991), *A Broadband Seismological Investigation of the 1989 Loma Prieta, California Earthquake: Evidence for Deep Slow Slip?* Bull. Seismol. Soc. Am. *81*, 1622–1646.

WATTS, A. B., WEISSEL, J. K., DUNCAN, R. A., and LARSON, R. L. (1988), *Origin of the Louisville Ridge and its Relationship to the Eltanin Fracture Zone System*, J. Geophys. Res. *93*, 3051–3077.

WIENS, D. A. (1989), *Bathymetric Effects on Body Waveforms from Shallow Subduction Zone Earthquakes and Application to Seismic Processes in the Kurile Trench*, J. Geophys. Res. *94*, 2955–2972.

WOODHOUSE, J. H., and DZIEWONSKI, A. M. (1984), *Mapping the Upper Mantle: Three-dimensional Modeling of Earth Structure by Inversion of Seismic Waveforms*, J. Geophys. Res. *89*, 5953–5986.

ZHANG, J., and KANAMORI, H. (1988a), *Depths of Large Earthquakes Determined from Long-period Rayleigh Waves*, J. Geophys. Res. *93*, 4850–4868.

ZHANG, J., and KANAMORI, H. (1988b), *Source Finiteness of Large Earthquakes Measured from Long-period Rayleigh Waves*, Phys. Earth Planet. Inter. *52*, 56–84.

ZHANG, J., and LAY, T. (1989a), *A New Method for Determining the Long-period Component of the Source Time Function of Large Earthquakes*, Geophys. Res. Lett. *16*, 275–278.

ZHANG, J., and LAY, T. (1989b), *Source Duration and Depth Extent of the June 22, 1977 Tonga Earthquake*, Bull. Seismol. Soc. Am. *79*, 51–66.

(Received September 12, 1992, revised/accepted March 12, 1993)

PAGEOPH, Vol. 140, No. 2 (1993)

0033–4553/93/020365–26$1.50 + 0.20/0
© 1993 Birkhäuser Verlag, Basel

Large Earthquake Doublets and Fault Plane Heterogeneity in the Northern Solomon Islands Subduction Zone

ZHENGYU XU[1] and SUSAN Y. SCHWARTZ[1]

Abstract—In the Solomon Islands and New Britain subduction zones, the largest earthquakes commonly occur as pairs with small separation in time, space and magnitude. This doublet behavior has been attributed to a pattern of fault plane heterogeneity consisting of closely spaced asperities such that the failure of one asperity triggers slip in adjacent asperities. We analyzed body waves of the January 31, 1974, $M_w = 7.3$, February 1, 1974, $M_w = 7.4$, July 20, 1975 (14:37), $M_w = 7.6$ and July 20, 1975 (19:45), $M_w = 7.3$ doublet events using an iterative, multiple station inversion technique to determine the spatio-temporal distribution of seismic moment release associated with these events. Although the 1974 doublet has smaller body wave moments than the 1975 events, their source histories are more complicated, lasting over 40 seconds and consisting of several subevents located near the epicentral regions. The second 1975 event is well modeled by a simple point source initiating at a depth of 15 km and rupturing an approximate 20 km region about the epicenter. The source history of the first 1975 event reveals a westerly propagating rupture, extending about 50 km from its hypocenter at a depth of 25 km. The asperities of the 1975 events are of comparable size and do not overlap one another, consistent with the asperity triggering hypothesis. The relatively large source areas and small seismic moments of the 1974 doublet events indicate failure of weaker portions of the fault plane in their epicentral regions. Variations in the "roughness" of the bathymetry of the subducting plate, accompanying subduction of the Woodlark Rise, may be responsible for changes in the mechanical properties of the plate interface.

To understand how variations in fault plane coupling and strength affect the interplate seismicity pattern, we relocated 85 underthrusting earthquakes in the northern Solomon Islands Arc since 1964. Relatively few smaller magnitude underthrusting events overlap the Solomon Islands doublet asperity regions, where fault coupling and strength are inferred to be the greatest. However, these asperity regions have been the sites of several previous earthquakes with $M_s \geq 7.0$. The source regions of the 1974 doublet events, which we infer to be mechanically weak, contain many smaller magnitude events but have not generated any other $M_s \geq 7.0$ earthquakes in the historic past. The central portion of the northern Solomon Islands Arc between the two largest doublet events in 1971 (studied in detail by SCHWARTZ *et al.*, 1989a) and 1975 contains the greatest number of smaller magnitude underthrusting earthquakes. The location of this small region sandwiched between two strongly coupled portions of the plate interface suggest that it may be the site of the next large northern Solomon Islands earthquake. However, this region has experienced no known earthquakes with $M_s \geq 7.0$ and may represent a relatively aseismic portion of the subduction zone.

Key words: Rupture process, fault plane heterogeneity, earthquake doublet, subduction zone, Solomon Islands.

---

[1] Institute of Tectonics and Earth Sciences Department, University of California, Santa Cruz, CA 95064, U.S.A.

Introduction

Large doublet or multiplet earthquakes separated closely in time, space and magnitude have occurred at several subduction zones including the Solomon Islands (LAY and KANAMORI, 1980; WESNOUSKY et al., 1986), New Hebrides (VIDALE and KANAMORI, 1983), Kermadec (CHRISTENSEN et al., 1991), and the Nankai trough (ANDO, 1975). Along both the northwest and southeast portions of the Solomon Islands as well as in the New Britain trench, doublet behavior is the common mode of occurrence for the largest earthquakes, making it distinctive from most other subduction zones. The mechanism responsible for earthquake triggering is not well understood, however the generation of compound earthquakes indicates heterogeneity in the faulting process. To enhance our knowledge of the role that fault plane heterogeneity plays in earthquake occurrence we investigate the rupture process of large doublet earthquakes and evaluate their influence on interplate seismicity patterns in the northern Solomon Islands Arc.

Three large doublet earthquakes occurred in 1971, 1974 and 1975 along the northern Solomon Islands Arc (Figure 1). LAY and KANAMORI (1980) analyzed the surface and body waves for all six events, constraining their focal mechanisms and seismic moments. Subsequently, the largest 1971 doublet (July 14, $M_w = 8.1$ and July 26, $M_w = 8.0$) was studied using more sophisticated body wave analysis (SCHWARTZ et al., 1989a). This paper will focus on the other two doublets (January 31, 1974, $M_w = 7.3$, February 1, 1974, $M_w = 7.4$, July 20, 1975, 14:37, $M_w = 7.6$ and July 20, 1975, 19:45, $M_w = 7.3$, hereafter 1974a,b, 1975a,b, respectively), applying modern body wave analysis to determine their rupture process.

The 1974 and 1975 earthquake pairs have northeast shallow dipping fault planes, which are consistent with an interpretation of underthrusting of the Solomon Sea Plate beneath the Ontong-Java Plateau (Figure 1). Subduction of the relatively high Woodlark Rise near 8° S separates the source areas of the 1974 and 1975 doublets and marks a transition in both the bathymetry of the trench and in the observed seismicity pattern. Northwest of the Woodlark Rise, the trench is deep (~ 8 km), linear and well defined and seismicity is abundant. To the south of this rise, the trench shallows, becomes undulatory, and seismicity decreases as the actively spreading Woodlark Ridge is subducted (Figure 1). We find that changes in the plate interface associated with subduction of the Woodlark Rise and/or Ridge have influenced the rupture process of the 1974 and 1975 doublet events.

Method and Data

Far-field P wave seismograms for a particular source-receiver pair can be modeled by the following equation (after KANAMORI and STEWART, 1976):

$$u(t) = (gc/4\pi\alpha^3\rho a)[R_p M_0(t) + R_{pP}A_{pP}M_0(t + \Delta t_{pP})$$
$$+ R_{sP}A_{sP}M_0(t + \Delta t_{sP})] * I(t) * Q(t) \tag{1}$$

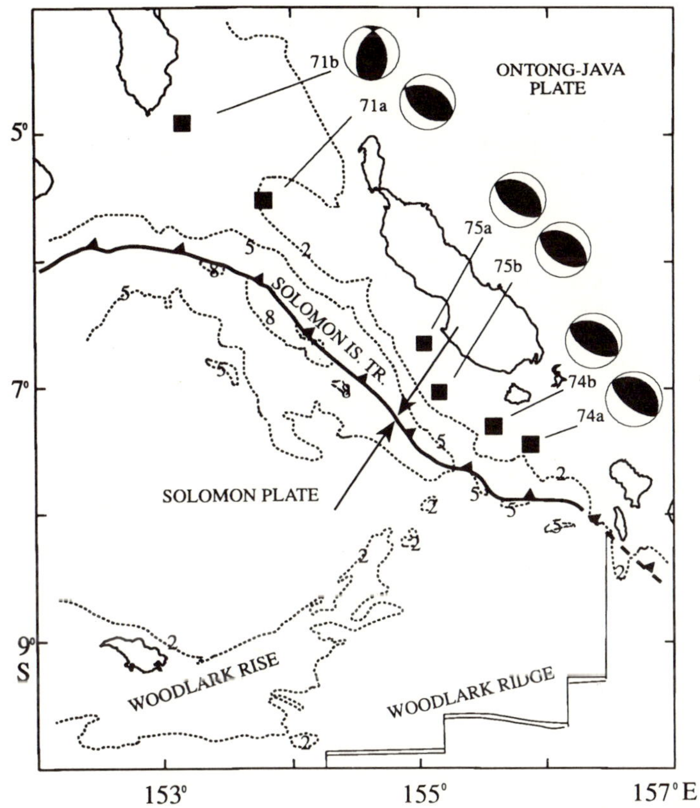

Figure 1

Map of the northern Solomon Islands region showing the locations of the six most recent large earthquake doublets. Lower hemisphere projections of the focal mechanisms are from LAY and KANAMORI (1980) except for the 1971a event that is from SCHWARTZ *et al.* (1989a). The arrows indicate relative motion vectors (WEISSEL *et al.*, 1982). Bathymetry in kilometers is indicated by the dashed lines.

where $u(t)$ is the far-field seismogram, g is geometric spreading, c is the free-surface effect at the receiver, α and ρ are P-wave velocity and density at the source, a is the radius of the earth, R accounts for the radiation pattern, A is the free-surface reflection coefficient, Δt is the travel time relative to P for the pP or sP phase, $M_0(t)$ is the source time function, $I(t)$ and $Q(t)$ are the instrument and attenuation impulse responses, and $*$ denotes convolution with respect to time. Large complex sources can be modeled as a sum of spatially and temporally distributed pulses by generalizing (1).

We invert P wave seismograms using an iterative multi-station deconvolution technique to determine the time, location and seismic moment of individual subevents of complex ruptures (KIKUCHI and KANAMORI, 1982, 1986). The multiple station inversion involves minimization of the squared difference between the observed and synthetic waveforms given by:

$$\Delta_k = \sum_{j=1}^{M} \int \left[x_j(t) - \sum_{k=1}^{N} m_k w_{jk}(t - t_k, f_k) \right]^2 dt$$

where x_j is the observed data at station j, m_k is the moment chosen for the kth iteration, w_{jk} is the synthetic wavelet at station j for iteration k, t_k is the timing of the kth subevent, f_k represents other source parameters, such as location, trapezoid shape, and focal mechanism and M is the number of stations. We will apply this analysis to the 1974 and 1975 Solomon Islands doublets.

In this paper, we use the terms, point, line and 2-D source to represent earthquake subevents that are constrained to one point, to lie along strike of the fault, or to lie along a two-dimensional grid both along strike and dip of the fault plane. We perform the inversion assuming each of the source parameterizations and after a number of iterations, if the residual waveform error is significantly decreased by increasing the dimension of the source model, we select the higher dimensional source. Otherwise, we use the source parameterization with the smallest degrees of freedom.

Our data consist of long-period vertical or horizontal component P wave seismograms recorded by the World Wide Standardized Seismograph Network (WWSSN). The parameters assumed in the construction of all Green's functions were P wave velocities $v_p = 6.0$ and 6.8 km/s at the surface and source respectively (S wave velocity is $v_p/1.732$), source density of 3.3 g/cm^3, water depth of 5 km, t^* of 1.0 s and geometric spreading factors appropriate for a Jeffreys-Bullen earth model. P wave onset times were picked from short-period data when available or from long-period data by correlating the first waveform cycles (especially, the first peaks) for stations nearby to one that had an accurate time determined from short-period data. The focal mechanisms are constrained to be those determined by LAY and KANAMORI (1980), dip = 28°, 34°, 36°, 40°, strike = 309°, 302°, 306°, 303°, slip = 90°, 90°, 90°, 90°, for the 1974a, 1974b, 1975a and 1975b earthquakes, respectively. We assume that the same trapezoidal source shape and fixed focal mechanism are appropriate for each subevent comprising the multiple rupture. Unreliable instrument magnifications for many of the WWSSN stations, as well as inadequacies in model Green's functions, make the use of absolute amplitudes in the simultaneous inversion difficult. Scaling factors were applied where necessary to bring the amplitude of the synthetic seismogram into better agreement with the observed. Stations for which scaling factors were necessary were excluded from determination of the body wave moment. Because the inversion technique inherently weighs seismograms with large amplitudes more than those with small amplitudes, we normalized all seismograms to a common gain before inversion.

Source Process of the 1974 and 1975 Doublet Events

1974a and 1974b Events

Seismograms from the 1974a and 1974b events are more complicated than the 1975 doublets events, although they are smaller. This may reflect heterogeneity in the mechanical properties of the fault plane. We find that a point source is adequate to model both the 1974a and 1974b events. To determine the best point source depth and source trapezoid, we computed residual waveform misfit (error) versus depth for a variety of trapezoid shapes of source time functions characterized by a rise time, t_0, and total duration, t_1 (Figure 2). The best variance reduction for the 1974a and 1974b events after 10 iterations yielded the misfit (error) values of 0.51 and 0.39, for depths at 27.5 km and 20.0 km, and source trapezoids of $(t_0, t_1) = (1, 5)$ and $(3, 8)$, respectively. Resulting source time functions, observed and synthetic waveforms are shown in Figures 3 (1974a) and 4 (1974b). Allowing point sources to be distribution along strike of the fault planes

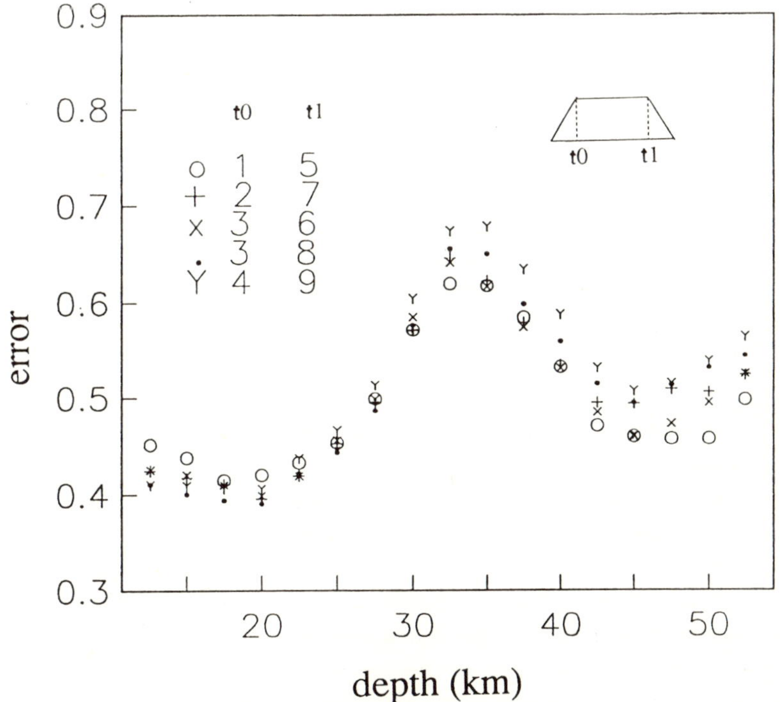

Figure 2

Residual waveform misfit (error) versus point source depth after 10 iterations of the simultaneous deconvolution of body waves for the February 1, 1974 (1974b) earthquake assuming various trapezoidal source time histories. The smallest waveform misfit is obtained for a depth of 20 km and is fairly insensitive to assumed trapezoidal shape.

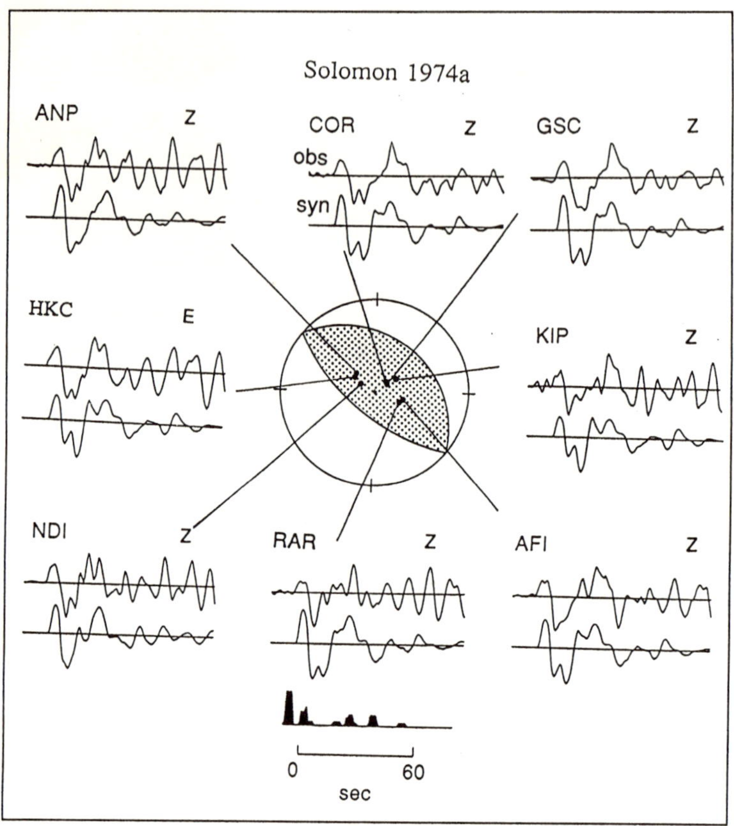

Figure 3

Observed (top) and synthetic seismograms (bottom) and the source time function resulting from a point source inversion of the January 31, 1974 (1974a) event waveforms. Lower hemisphere focal mechanism is from LAY and KANAMORI (1980). Station name and component are also indicated.

did not substantially reduce waveform misfit for either the 1974a or 1974b events. Although both of these earthquakes are adequately modeled as point sources, we believe that they ruptured small yet finite portions of the fault plane, the small extent of fault rupture is not discernible through analysis of long-period body waves.

The source time function for the 1974a event (Figure 3) consists of two main pulses of moment release, each with an approximate 5 s duration. Although there is some moment released later in time, the latter parts of the seismograms are not fit well enough to establish its timing. To estimate a rupture area for this event we assume that the two episodes of moment release rupture distinct circular regions of the fault plane and use an average rupture velocity of 2.5 km/s to obtain rupture radii of 13 km for each subevent. Applying the same approach to the four 8 s pulses of moment release associated with the 1974b event (Figure 4) yields

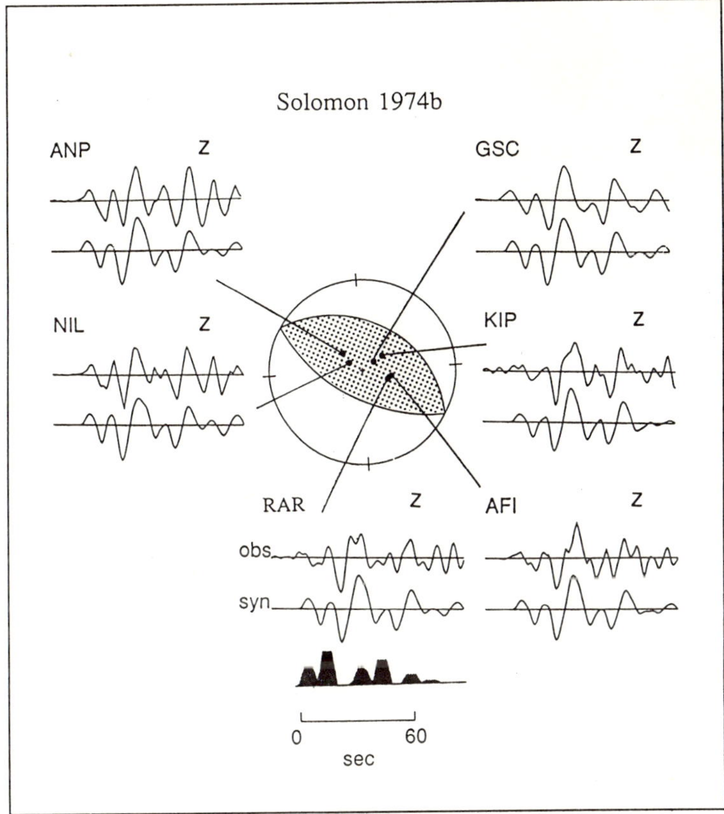

Figure 4

Same as Figure 3 for the February 1, 1974 earthquake (1974b).

rupture radii of 20 km for each of its four main subevents. These events have body wave moments of 0.3 and 0.9×10^{27} dyne-cm for the 1974a and b events, respectively.

1975b Event

Waveforms from the 1975b event were also adequately fit, assuming a point source. Residual waveform misfit (error) versus depth after 10 iterations of the simultaneous deconvolution yields a minimum error of 0.27 with $t_0 = 3$, $t_1 = 8$, and depth = 15 km. The resulting source time function and synthetic and observed waveforms are shown in Figure 5 for 8 of the 12 stations used in the inversion. The fit of the synthetics to the observed waveforms is quite good, revealing a very simple source process. One subevent with an 8 s duration dominates the rupture process (Figure 5). Because of the simple source, the rupture length of this event can be

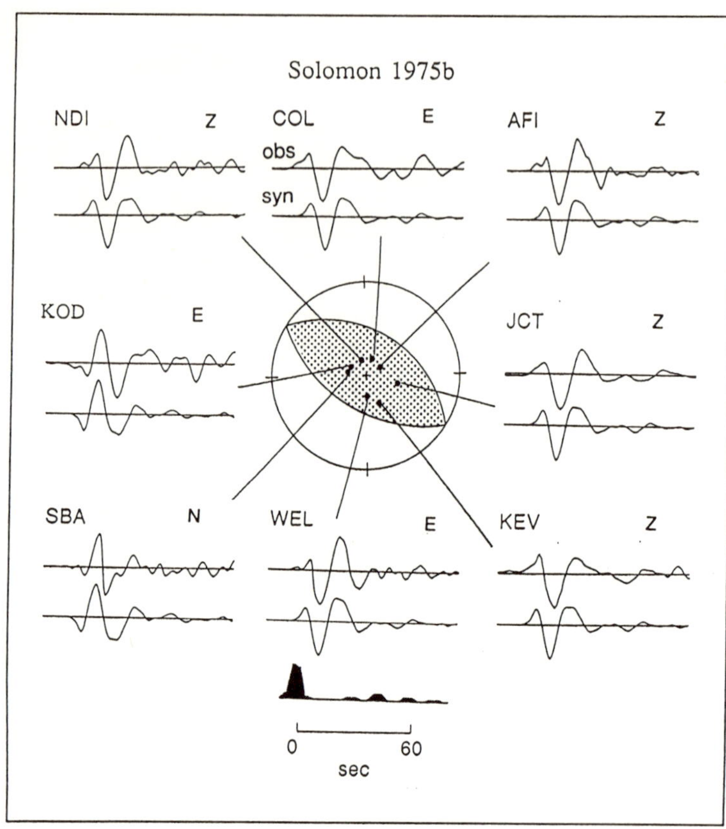

Figure 5
Same as Figure 3 for the July 20, 1975, 19:45 event (1975b), for 8 of 12 stations.

more confidently approximated by its source duration multiplied by a reasonable rupture velocity than the 1974 events. Using a source duration of 8 s and a rupture velocity of 2.5 km/s, we estimate the rupture length to be about 20 km. With no observable directivity, we make the conservative assumption that the rupture propagated circularly near the epicenter. An average body wave moment of 1.5×10^{27} dyne-cm was obtained. This very simple source for an $M_w = 7.3$ earthquake is quite striking in comparison with the complexity of the somewhat smaller 1974 doublet events.

1975a Event

Both a point source and line source parameterization are inadequate to fit P waveforms from the 1975a event. We performed the inversion specifying point source locations on a two-dimensional grid consisting of 11 sources located along

150 km in the strike direction and 5 sources located along 60 km in the dip direction. We determined the depth of rupture initiation by performing one iteration of the inversion using only the first cycle of the waveforms. We then used this initial depth to solve for the time and position, along the strike and dip of the fault plane, of multiple subevents relative to the hypocenter. Several inversions were performed, assuming various shapes for the trapezoidal time history. The largest error reduction was obtained using an initial source depth of 25 km and subevent source trapezoids with $t_0 = 4$ s and $t_1 = 9$ s. The observed and synthetic waveforms and source time functions are plotted for 8 of 17 stations in Figure 6. The source time functions consist of two pulses of moment release with the second pulse more than twice as large as the first. The peaks of these two major episodes of moment release are closer together at the western stations than at the eastern stations, indicating rupture propagation towards the west. However, the duration of the

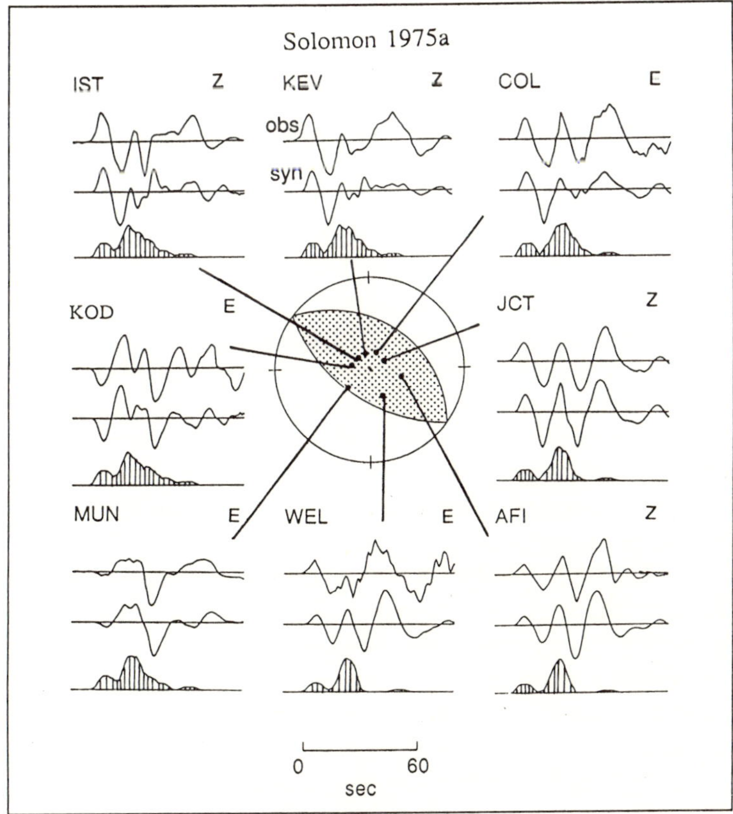

Figure 6

Observed and synthetic seismograms and deconvolved source time functions resulting from the two-dimensional simultaneous inversion of waveforms from the July 20, 1975, 14:37 event (1975a), for 8 of 17 stations.

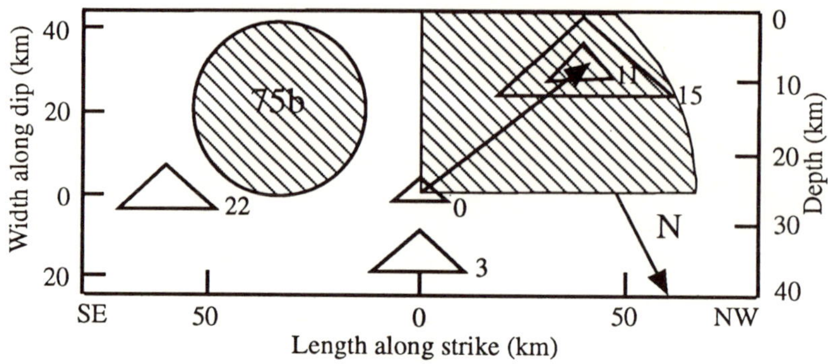

Figure 7

Spatial distribution of the five largest subevents for the 1975a earthquake. The size of the triangles is proportional to the seismic moment of each subevent and the numbers indicate their onset times. Over half of the moment was released during two subevents west of the epicenter and we interpret the region between the epicenter and these subevents as the dominant asperity. Also shown is the source region of the second doublet (75b) which occurred only 5 hours after the first.

second major peak of moment release is longer at stations to the north and northwest, indicating that rupture also propagated towards the southeast. Figure 7 shows the spatial distribution of the five largest subevents with the remaining smaller subevents distributed over the fault area. The largest subevent occurred updip and to the west of the hypocenter, 15 s after rupture initiation. Most of the moment release is located within 100 km along strike and 50 km along dip of the hypocenter, with no subevents deeper than 35 km.

The body wave moment is 2.2×10^{27} dyne-cm, of which more than half is contained in what we define as the 1975a main asperity (shaded region in Figure 7). The asperity area is approximated as a truncated quarter circle centered on the azimuth from the hypocenter to the largest subevent and rupturing to the surface. Figure 7 also shows the approximate rupture area of the 1975b event discussed in the last section. Its single source rupture process and proximity to the main 1975a asperity make it appear to be the 1975a event's 'five-hour later subevent'. The nearly identical focal mechanisms for the two 1975 doublet events (LAY and KANAMORI, 1980) support the idea that both occur on a single fault plane. The 1975b event appears to fill a gap in the subevents of the 1975a event. The close but distinct positions of the main 1975 doublet asperities support the hypothesis of LAY and KANAMORI (1980) that earthquake doublets in the Solomon Islands result from slip on one asperity triggering slip in a closely adjacent asperity. Although it is more difficult to identify distinct asperities associated with the rupture of the 1974 doublet events, their source regions do not appear to overlap significantly, also consistent with the asperity triggering model. We note that the latest subevent of the 1975a event occurred towards the southeast, close to the 1975b asperity, the second event of the doublet, possibly loading that section of the fault plane. A

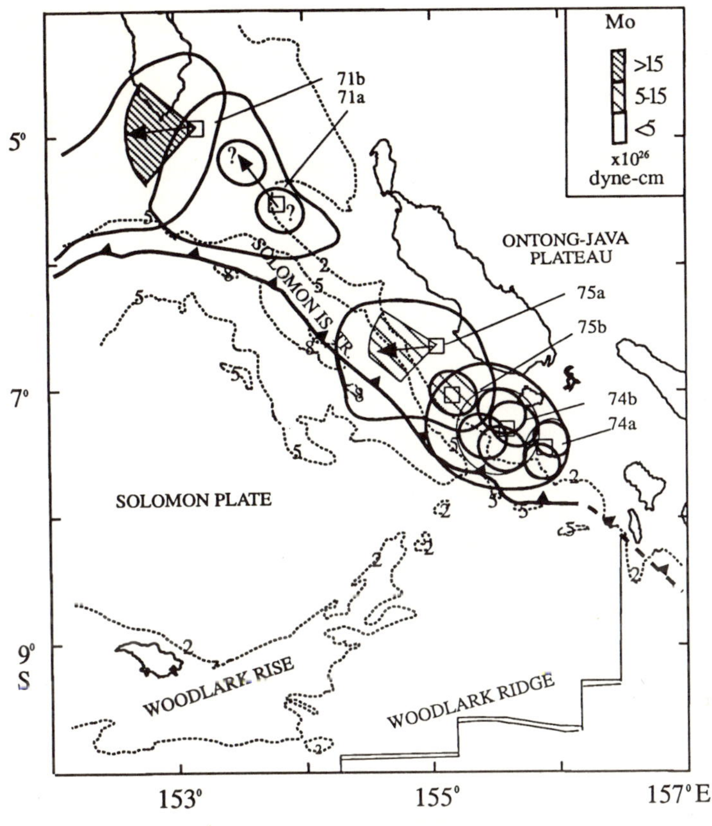

Figure 8

Schematic map indicating approximate body wave source areas and moments listed in Table 1 for the 1974 and 1975 doublets determined in this study and the 1971 doublet from SCHWARTZ *et al.* (1989a). The arrows indicate the rupture direction and extent for events where source directivity was observable. Two-day aftershock areas are indicated with heavy lines (1971b has only a portion of the aftershock area shown).

similar phenomenon also occurred during the 1971 doublet events (SCHWARTZ *et al.*, 1989a) where the latter subevents of the first doublet event (1971a) rupture near the initiation point of the second doublet event (1971b), efficiently loading the portion of the fault plane soon to fail (Figure 8).

Fault Plane Heterogeneity

Fault Plane Coupling and Strength Inferred from Rupture Process

Our study of the source process of the 1974 and 1975 Solomon Islands doublet earthquakes has resulted in estimates of their body wave moments and rupture areas. These values are shown in Figure 8 and listed in Table 1 along with those

Table 1

Earthquake source parameters

| Event | | | M_s | M_{0s} | M_{0b} | Main Asperities | | |
| yr | mo | day | | | | M_{0b1} | Area | Stress |
					(exp 26 dyne-cm)		(km × km)	Drop (bar)
1971	7	14	7.9	120	30			
1971	7	26	7.9	180	80	45	3400	54
1974	1	31	7.0	10	3	1 × 2 #	490	22
1974	2	1	7.1	14	9	2 × 4 #	1300	10
1975	7	20	7.9	34	22	13	2400	27
1975	7	20	7.7	12	15	8	1300	41

M_{0s} is surface wave moment from LAY and KANAMORI (1980).

M_{0b} and M_{0b1} are body wave moment from SCHWARTZ *et al.* (1989a) and this study.

#: ×2 and ×4 represent the number of similar size asperities.

determined for the second 1971 doublet event by SCHWARTZ *et al.* (1989a). The spatial distribution of moment release associated with the first 1971 doublet event was not well resolved (SCHWARTZ *et al.*, 1989a); therefore values for the body wave source dimensions have been omitted from Table 1 and are shown only schematically in Figure 8. Also shown in Figure 8 are the two-day aftershock areas for all of the doublet events. For the largest doublet in 1971, the aftershock areas are much larger than the body wave source areas. These earthquakes also have a large discrepancy between the surface and body wave moments (LAY and KANAMORI, 1980; Table 1). The body wave source regions are a considerably larger proportion of the aftershock areas for the smaller 1974 and 1975 doublet events; however, for some of the events, substantial discrepancies still exist between their body wave and surface wave moments. The large difference between body wave and surface wave moments for all but the 1975b event requires a long-period component of moment release over their entire aftershock areas. The limited band width of the body waves precludes retrieval of the total moment. In this paper we limit ourselves to the higher-frequency characteristics of the rupture process.

For subduction zone earthquakes, RUFF and KANAMORI (1980) have associated earthquake size and the degree of coupling along the plate boundary, where larger earthquakes indicate stronger coupling. Along the northern Solomon Islands subduction zone, the sizes of the largest earthquakes, as determined by their surface wave moments, decrease from north to south, suggesting a decrease in plate coupling. The body wave seismic moments of the six largest earthquakes also show this decreasing trend (Table 1). We further investigate variations in plate coupling along this boundary by identifying the regions radiating most of the body wave moment as asperities and differentiating relative degrees of coupling or fault strength between asperities by comparing static stress drops associated with their failure. It is well known that in general, stress drops of interplate earthquakes are smaller than those of intraplate events (KANAMORI and ANDERSON, 1975). This has been interpreted in terms of lower strength on plate boundary faults compared to faults within plates since frictional strength at the boundary of two solids having frictional movement is generally less than the shear strength within the solids. In an analogous manner, we assume that variations in stress drop determined for the Solomon Islands earthquake asperities reflect differences in the mechanical strength of the fault plane.

We calculate asperity stress drops using the expression given by KANAMORI and ANDERSON (1975) for a circular fault embedded in a whole space, $\Delta\sigma = 2.4\, M_0/S^{1.5}$, where M_0 is the seismic moment associated with the asperity (M_{0b1} in Table 1) and S is the asperity area. Due to large uncertainties in our estimates of asperity area and moment, resulting values for asperity stress drop (Table 1) are crude at best. However, the observed north to south decreasing trend for asperity stress drop is similar to the pattern apparent in both the surface and body wave seismic moment and supports our interpretation of a north to south decrease in plate coupling. The asperity region associated with the largest seismic moment, the 1971b epicentral asperity, failed with the highest stress drop, the 1975a and b asperities failed with intermediate stress drops, while the 1974 doublet source regions, which occupy most of their aftershock area and released the smallest seismic moments, failed with the lowest stress drop.

The relatively low moment release and large source areas of the 1974 doublet events are inconsistent with the presence of asperities, making this region of the fault plane anomalous with respect to the northern portion of the arc (Table 1 and Figure 8). A progression in stress drop and inferred fault strength is apparent along the northern Solomon Islands Arc with the strongest plate coupling at the junction of the Solomon Islands and New Britain trenches, becoming weaker to the south as the intersection between the Solomon trench and the Woodlark Ridge is approached. The reduction in coupling and fault strength and disappearance of coherent asperities in the source region of the 1974 doublet events may be related to subduction of the Woodlark Rise and/or Woodlark Ridge (Figure 8). Coincident with subduction of the Woodlark Rise, the Solomon Islands trench shoals, becomes discontinuous and undulatory, clearly indicating disruption of the smoother mode of subduction to the north. The Woodlark Ridge is the site of active spreading and exhibits a few unique geologic effects often associated with ridge subduction such as: increased and anamolous arc magmatism, complex differential vertical movements, shallowing of the trench axis, and reduction in seismicity (TAYLOR and EXON, 1987). These effects clearly indicate that subduction is strongly influenced by introduction of the Woodlark Ridge into the trench. It seems reasonable that in addition to these effects, elevated temperatures in the vicinity of the ridge would result in reduced coupling between the subducting and overriding plates from the Woodlark Rise, the boundary of the spreading region, to the southeast. Relative high heat flow values of 124 mW/m^2 have been measured on the Woodlark Rise (HALUNEN and VON HERZEN, 1973), and may be responsible for the reduced coupling in this region. The 1974b event has the lowest stress drop of the Solomon Islands doublets and is located at the intersection of the Woodlark Rise and the trench. Figure 8 provides a crude estimate of fault coupling and strength along the plate interface in the northern Solomon Islands Arc; next we investigate how this coupling and strength variation affect the interplate seismicity pattern.

Relocation of Underthrusting Seismicity

All earthquakes with documented underthrusting focal mechanisms that occurred from 1964 to 1990 in the Solomon Islands Arc north of the Woodlark Ridge (between 4° and 8° S and 152.5° and 156.2° E) were relocated using ISC arrival time data and the method of joint hypocentral determination (JHD) described by DEWEY (1972). Focal mechanisms for events in the Solomon Islands before 1977 were determined by COOPER (1985), JOHNSON and MOLNAR (1972), PASCAL (1979), and RIPPER (1975). These mechanisms and Harvard Centroid Moment Tensor (CMT) solutions for events after 1977 were examined. All events having northwest striking, southwest shallow (99% of dips are $\leq 50°$) dipping nodal planes, and focal depths less than 100 km were considered to be underthrusting earthquakes.

Our application of the JHD method involves the location of a small number of widely distributed events with respect to a calibration event to determine station corrections and variances for nearly 100 stations that are then used in a single event location program to calculate hypocenters and error estimates for all events. The JHD station corrections account for travel time biases resulting from near-source velocity anomalies. To assure that our source region is sufficiently small so that station corrections adequately correct for near-source structure, we divide the northern Solomon Islands Arc into two segments (Figure 9) and calculate separate station corrections for each segment. The same calibration event is used to calculate station corrections in both segments in order to link them. To obtain a more accurate location for the calibration event, on which all other earthquake locations depend, its epicenter was relocated using JHD and calibration station corrections rather than a calibration event. Azimuthally dependent station corrections of DZIEWONSKI and ANDERSON (1983) for 29 well-distributed stations were selected to minimize travel time bias arising from velocity anomalies beneath the stations. The calibration station corrections do not account for travel time due to near-source anomalies. Table 2 lists all the events that were relocated; the event on July 23, 1984 was used as the calibration event and its depth was fixed to the Harvard CMT depth. Since the magnitude of the calibration event is only M_s 5.7, the CMT depth and focal depth should be equivalent.

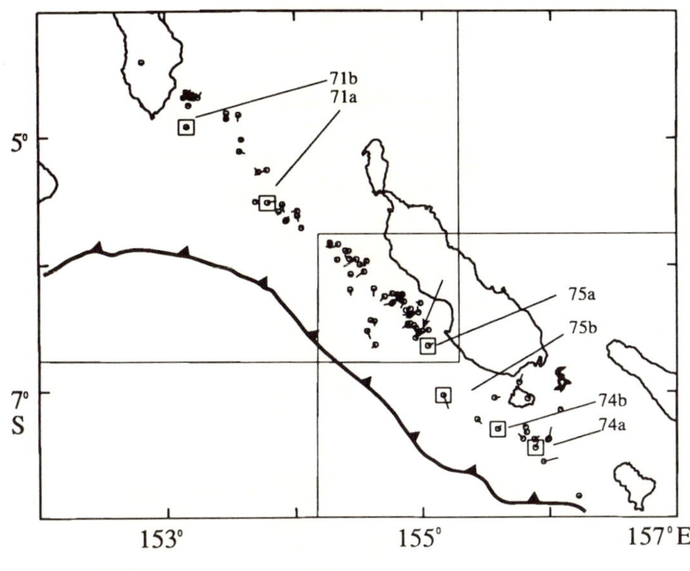

Figure 9

Relocated (circles) and ISC (tails) epicenters of 85 underthrusting earthquakes in the northern Solomon Islands Arc. The six large doublet events are labeled and marked with small boxes. The two large boxes indicate the subdivision of the region used in the relocation and the arrow marks the location of the master event to which all earthquake locations are tied. No systematic shift in location is evident after relocation and few events moved more than 10 km.

After relocation, 48 of the 85 events had epicentral and hypocentral depth errors of less than 8 km and 15 km respectively, 19 events had epicentral errors less than 10 km and hypocentral depth errors between 15 and 20 km, and the remaining events had errors greater than these. Figure 9 shows both the ISC (tail) and relocated (circle) epicenters for all events. The average change in epicentral location is less than 10 km, with no systematic shift in position after relocation. Figure 10 shows the ISC and relocated events in cross section. The northeast dipping plate interface is more clearly defined after relocation, especially in the shallow portion above 30 km. Although definition of the subducted lithosphere is improved after relocation, significant scatter still exists. Some of this scatter may be attributed to the relatively large along strike distance (~ 500 km) collapsed onto one cross section. The plate geometry may vary along strike between the Solomon–New Britain corner and the Woodlark Rise, causing the observed scatter. Focal depths shallow considerably after relocation with most events confined to depths above 65 km. Although 65 km seems deep for the transition from unstable slip to stable slip, ZHANG and SCHWARTZ (1992) found a cutoff depth of about 70 km for underthrusting earthquakes in the northern Solomon Islands (New Ireland) subduction zone. This subduction zone possessed the deepest underthrusting earthquakes which ZHANG and SCHWARTZ (1992) attributed to colder temperatures at

Table 2

Relocated events

| Date | | | Time UT | Lat. | Lon. | Depth | M_s | Strike | Dip | Rake | Mechanism |
yr	mo	day	h m sec	N deg	E deg	km	*	deg	deg	deg	Ref.
1964	3	6	1857:15.9	−5.97	154.49	63.6	4.8	312	45	90	JM
1964	4	17	6 0: 0.1	−6.59	154.95	68.7	4.6	280	46	31	R
1967	4	10	459:53.9	−7.39	155.79	23.5	4.5	307	30	74	JM
1967	4	10	15 2:43.3	−7.30	155.82	23.4	5.0	290	21	77	JM
1967	10	4	1721:18.2	−5.67	153.93	16.6	5.2	305	50	90	JM
1967	10	8	18 8:17.6	−5.59	154.03	44.0	4.5	354	44	37	R
1967	12	25	123:34.0	−5.28	153.73	45.5	7.2	292	44	90	JM
1969	8	5	1632:25.7	−5.27	153.79	48.2	4.8	312	50	90	P
1970	3	28	745:58.3	−6.20	154.62	31.2	5.2	314	50	46	PC
1970	9	23	12 4:52.4	−6.46	154.63	11.4	5.2	304	45	90	PC
1971	7	14	611:31.6	−5.53	153.80	53.6	7.9	295	45	80	SH
1971	7	26	023:20.4	−4.93	153.17	23.9	7.9	240	40	55	LK
1971	8	9	1212: 1.6	−5.85	154.35	41.9	5.4	308	49	90	PC
1971	12	4	225:51.3	−5.99	154.57	56.5	5.4	299	54	90	PC
1971	12	11	725:18.6	−6.07	154.55	39.3	4.8	338	40	49	PC
1974	1	31	2330: 6.6	−7.46	155.89	29.9	7.0	309	28	90	LK
1974	2	1	312:32.5	−7.31	155.60	12.5	7.1	302	34	90	LK
1975	7	20	1437:41.5	−6.66	155.05	44.5	7.9	306	36	90	LK
1975	7	20	1954:28.0	−7.04	155.17	29.4	7.7	303	40	90	LK
1975	8	21	946:43.3	−6.56	154.98	39.2	4.8	333	30	90	PC
1976	2	22	1829: 0.2	−6.30	154.86	54.1	5.2	334	50	90	PC
1977	8	6	1126:13.2	−7.07	155.83	71.4	5.5	289	47	63	HVD
1977	8	12	0 7:53.2	−6.53	155.05	51.9	5.7	306	42	75	HVD
1977	11	7	2257:17.1	−7.16	156.09	68.8	5.7	309	44	112	HVD
1978	3	31	7 1:57.1	−6.48	154.89	36.7	5.4	311	39	84	HVD
1978	9	20	1520:25.3	−5.52	153.70	19.8	4.6	329	34	122	HVD
1979	8	13	3 3:46.5	−4.87	153.48	54.6	6.0	294	23	33	HVD
X1980	2	6	550:57.8	−7.87	156.22	47.0	5.6	345	31	109	HVD
1980	2	12	320:21.3	−4.70	153.22	37.0	6.4	306	35	128	HVD
1980	4	6	2126:16.5	−6.39	154.98	63.6	5.2	299	44	79	HVD

1980	5	5	2259:55.7	−5.90	154.40	40.7	5.8	319	44	88	HVD
X1980	5	10	146:25.0	−5.91	154.43	62.0	6.2	317	41	94	HVD
X1980	5	14	1126: 1.9	−6.01	154.51	48.5	6.3	322	41	81	HVD
1980	5	15	2154:40.1	−5.97	154.43	55.8	5.8	328	42	75	HVD
1980	7	21	045: 9.8	−6.21	154.44	30.0	5.7	333	36	108	HVD
X1980	9	28	1825:59.9	−6.27	154.83	48.1	5.9	311	43	94	HVD
1980	10	20	1054:44.0	−6.25	154.85	63.9	5.5	309	39	101	HVD
1981	3	6	1718:55.5	−6.25	154.85	58.9	5.4	320	46	69	HVD
1981	4	23	1215:53.0	−6.36	154.91	64.2	5.6	325	44	83	HVD
1981	7	4	728: 9.6	−7.39	155.99	52.8	5.4	293	32	67	HVD
1981	7	13	1440: 7.4	−7.39	156.00	33.4	5.4	312	32	90	HVD
1981	7	30	1511:15.0	−5.86	154.28	47.2	5.3	297	26	144	HVD
1981	10	7	832:58.1	−6.41	154.90	49.0	5.5	318	40	104	HVD
1981	11	25	1040:57.8	−4.82	153.48	49.7	5.4	294	36	30	HVD
1981	12	13	139:17.0	−6.40	154.92	51.9	5.8	312	37	94	HVD
1981	12	13	1324:20.6	−6.37	154.88	56.5	5.4	307	35	88	HVD
1982	9	29	2046:55.1	−7.33	155.82	60.4	5.0	318	32	71	HVD
1983	2	9	7 2:36.9	−7.57	155.96	65.3	5.7	291	25	75	HVD
X1983	3	18	9 5:49.4	−4.84	153.57	63.8	7.8	308	49	120	HVD
X1983	3	20	1345:48.1	−4.68	153.21	55.2	6.3	316	34	134	HVD
1983	3	23	6 9:27.9	−6.45	154.60	13.6	6.2	328	34	105	HVD
1983	3	23	826:53.6	−6.54	154.57	6.8	5.6	319	41	30	HVD
1983	3	23	14 4:46.1	−6.64	154.64	18.0	5.4	287	36	61	HVD
1983	3	24	1047:40.9	−4.70	153.26	58.2	5.5	315	38	145	HVD
1983	4	12	2244:50.9	−5.12	153.58	41.9	5.5	335	37	115	HVD
X1983	4	15	444: 4.4	−6.49	154.94	42.5	5.5	312	35	86	HVD
X1984	7	5	521:51.8	−6.09	154.44	42.4	6.5	320	44	98	HVD
1984	7	10	1812:23.1	−6.31	154.77	53.7	5.5	305	42	74	HVD
#1984	7	23	6 6: 7.0	−6.54	155.00	47.6	5.7	307	38	73	HVD
1984	7	26	1835:52.7	−6.24	154.81	56.0	5.3	319	44	90	HVD
1984	11	22	1437:39.4	−6.29	154.84	54.3	5.8	313	45	97	HVD
1984	11	26	725:44.4	−6.24	154.77	55.4	5.5	314	42	95	HVD
1984	12	16	1951:55.6	−4.68	153.18	56.7	5.9	328	39	145	HVD
1984	12	21	1437:18.9	−4.76	153.18	60.3	4.9	279	11	86	HVD
1985	4	8	1617:13.9	−5.73	154.06	37.6	5.6	314	42	89	HVD
1985	5	4	0 4:23.3	−6.31	154.99	59.9	5.0	291	49	121	HVD

Table 2 (*Contd*)

Relocated events

Date			Time *UT*	Lat.	Lon.	Depth	M_s	Strike	Dip	Rake	Mechanism
yr	mo	day	h m sec	N deg	E deg	km	*	deg	deg	deg	Ref.
1985	5	24	515:31.4	−6.26	154.82	55.5	5.1	335	44	93	HVD
1985	6	5	23 4:56.5	−4.66	153.16	56.6	6.3	298	34	118	HVD
1985	6	27	1246: 1.0	−6.54	154.97	44.3	5.2	320	34	100	HVD
X1985	7	3	436:52.6	−4.41	152.82	24.8	7.2	329	37	106	HVD
1985	12	28	1028:46.0	−5.84	154.28	41.9	5.2	307	43	113	HVD
X1986	3	13	536:32.3	−7.07	155.57	79.3	6.1	295	46	62	HVD
1986	4	5	229:16.1	−6.94	155.77	67.1	5.2	294	44	118	HVD
1986	9	14	2058:24.2	−6.47	154.91	46.1	6.1	290	35	65	HVD
X1986	10	14	1653:11.0	−5.03	153.60	46.9	6.6	327	40	80	HVD
X1986	10	24	258:47.8	−5.60	153.88	38.8	6.3	329	36	110	HVD
1986	10	25	435:53.7	−5.63	154.02	34.7	5.3	325	40	87	HVD
1986	11	10	2152:46.6	−7.39	155.88	58.6	5.4	290	28	68	HVD
1987	7	20	2154:38.9	−6.41	154.91	45.8	5.3	326	39	102	HVD
1987	8	11	1744:53.3	−6.32	154.76	57.4	5.6	308	39	85	HVD
1987	8	21	1944:19.6	−6.51	154.96	48.4	5.6	323	39	101	HVD
1987	9	15	850:11.3	−4.70	153.14	55.6	6.0	307	32	131	HVD
1988	7	8	1635:36.5	−6.26	154.71	63.2	5.6	324	43	95	HVD
1990	7	7	1627:51.3	−5.97	154.34	45.5	5.1	285	28	130	HVD
1990	10	23	1514:28.7	−5.55	153.91	33.2	4.8	326	37	104	HVD

X represents events used to obtain station corrections.

\# represents the calibration event.

*M_s: For larger events, M_s values were from NEIC, while the 70s doublets were from EDR.

For smaller events, M_s values were derived from M_0 (CMT) using the relationship, $M_s = (2/3) \log M_0$ (KANAMORI and ANDERSON, 1975), since 1977; or from ISC's m_b, $M_s = 1.59 m_b - 3.97$ (RICHTER, 1958), before 1977.

Mechanism Reference: JM: JOHNSON and MOLNAR (1972), R: RIPPER (1975), P: PASCAL (1979), PC: COOPER (1985), SH: SCHWARTZ *et al.* (1989a), LK: LAY and KANAMORI (1980), HVD: Harvard CMT.

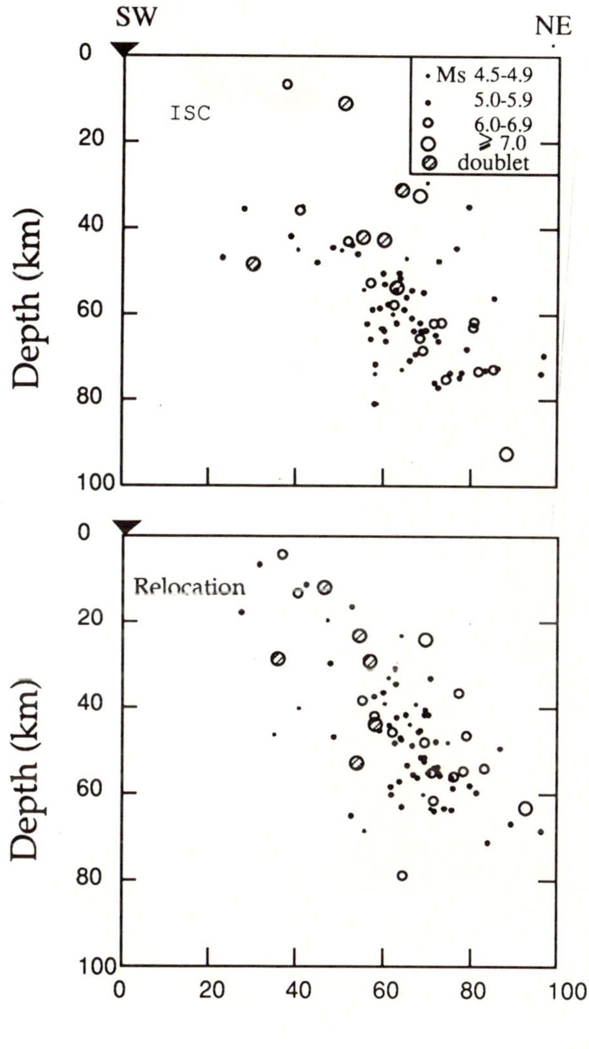

Figure 10

Cross sections perpendicular to the trench (marked by arrow) of ISC (top) and relocated (bottom) hypocenters. The shallow portion of the thrust interface is more clearly defined by the relocated hypocenters. The maximum depth of underthrusting seismicity shallows after relocation.

the deep portions of the thrust contact due to a fast convergence velocity quickly transporting cold slab material to depth. The persistence of deep underthrusting earthquakes in the northern Solomon Islands after relocation suggests that it is a real feature of this subduction zone and not an artifact of near-source velocity structure.

Fault Plane Coupling, Strength and Interplate Seismicity

Figure 11 shows the relocated underthrusting earthquakes superimposed on our crude map of fault strength. Few earthquakes occur within the better defined asperity regions, while most events locate between the 1971a and 1975a asperities. The observation that asperity regions remain locked between earthquake cycles has been made at several different subduction zones including the Aleutians (HOUSTON and ENGDAHL, 1989) and the Kurile Islands (SCHWARTZ *et al.*, 1986b) as well as in continental faulting environments (MENDOZA and HARTZELL, 1988; OPPEN-HEIMER *et al.*, 1990; HARTZELL and IIDA, 1990).

Figure 12 shows all events with $M_s \geq 7.0$ near the study region since 1900. Focal mechanisms have not been determined for the pre-1967 earthquakes, however the location of these events within the seismogenic zone defined by the more recent underthrusting seismicity suggests that they also occurred on the thrust interface.

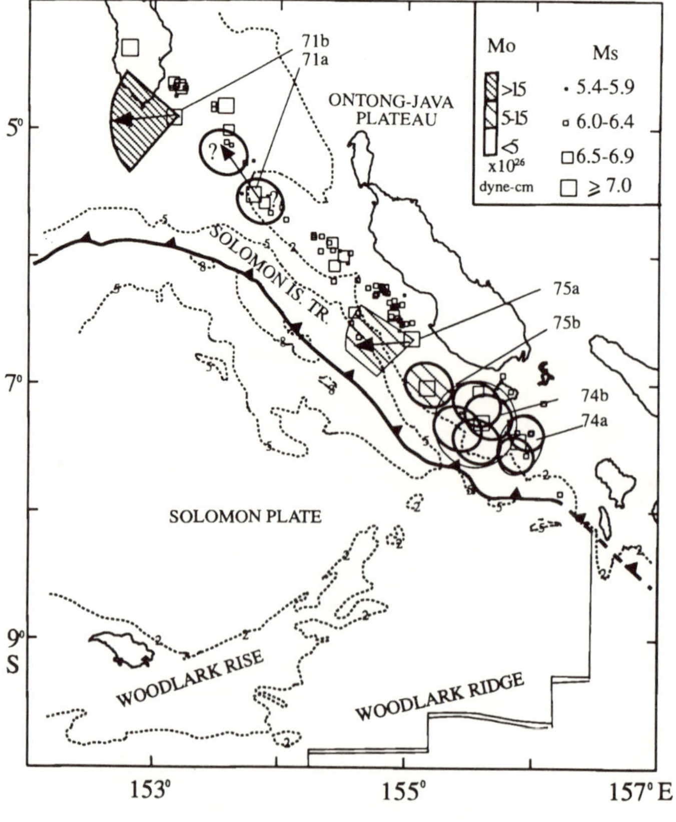

Figure 11

Relocated underthrusting seismicity (boxes) superimposed on the fault asperity map of Figure 8. Small magnitude events tend to occur outside the strongest portions of the plate interface (darkly hatchured areas) and concentrate in the region between the 1971a and 1975a doublet events.

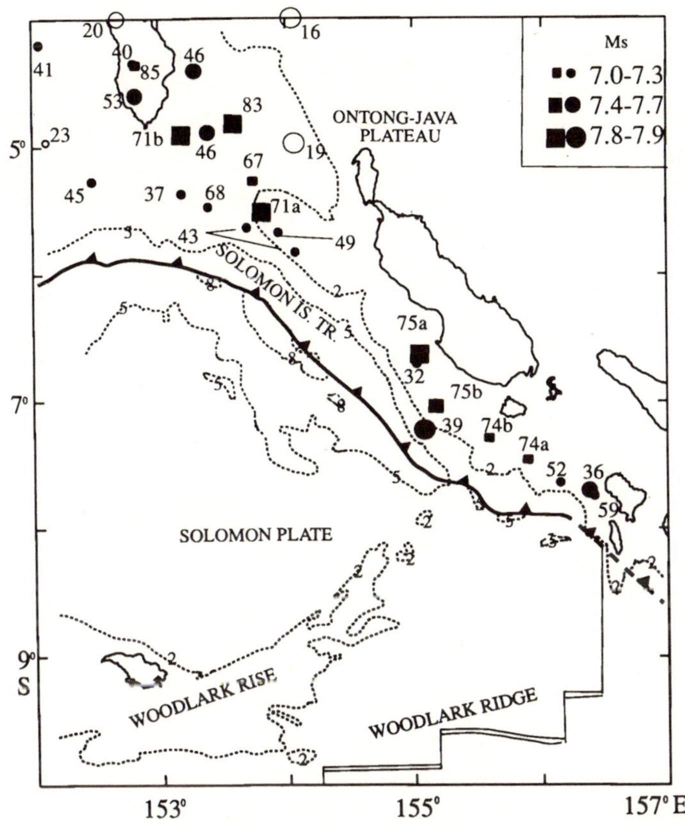

Figure 12

Large earthquakes ($M_s \geq 7.0$) indicated by year (two digit numbers) and M_s ranges since 1900. Solid squares are recent events since 1967 that have been relocated in this study, solid circles are early events relocated by McCann (1980), and open circles are historic events from Gutenberg and Richter (1954). The arc segment between the 1971a and 1975a events has experienced few known $M_s \geq 7.0$ earthquakes in historic times suggesting that subduction is occurring aseismically.

The historic record reveals that the asperity regions have experienced other large ($M_s \geq 7.0$) earthquakes in the recent past. For example, the 1943 and 1949 earthquakes with $M_s = 7.3$ and 7.1 (Abe, 1981) may have ruptured the 1971a asperities, and the 1932 and 1939 earthquakes with $M_s = 7.1$ and 7.8 (Abe, 1981) may have ruptured the 1975a and b asperities. We have inferred that the source region of the 1974 doublet events is weak compared with the source regions of the 1971 and 1975 doublets. This area has some small magnitude events, but contains no $M_s \geq 7.0$ earthquakes beside the 1974 doublet. In a previous section, we postulated that the change in fault coupling and strength may be due to the proximity of the 1974 source region to the actively spreading Woodlark Ridge and Woodlark Rise. Although three earthquakes in 1936 ($M_s = 7.4$), 1952 ($M_s = 7.2$), and 1959 ($M_s = 7.1$) ruptured portions of the fault plane closer to the Woodlark Ridge than the 1974 doublet, there is an unexpected relatively low heat flow of 58 mW/m^2 just seaward of the trench near these three events (HALUNEN and VON HERZEN, 1973). This low heat flow may allow locally isolated regions of strong plate coupling.

The most enigmatic portion of the northern Solomon Islands plate interface lies between the 1971a and 1975a earthquakes. This region fails frequently in small magnitude events (Figure 11), but has not generated any known $M_s \geq 7.0$ earthquakes in the historic past (Figure 12). The absence of large interplate slip over this 120 km span suggests that subduction is occurring aseismically. However, the position of this segment sandwiched between two strongly coupled portions of the plate interface and the high level of seismicity at its downdip edge may indicate that this region is presently locked and may be the site of the next large Solomon Islands earthquake. Studies of the stress state in subducted lithosphere (CHRISTENSEN and RUFF, 1983, 1988; DMOWSKA and LOVISON, 1988, 1992) have suggested that stresses in the outer rise are influenced by changes in stress on the thrust interface and therefore may be used to identify regions with high potential for future large underthrusting earthquakes. These studies have shown that outer rise compressional events either precede large underthrusting earthquakes or occur in regions that have been identified as present seismic gaps. In these cases the thrust interface is locked and compressional stresses can build in the outer rise. Tensional outer rise events tend to follow large underthrusting earthquakes in response to slab-pull forces. Evidence against the hypothesis that the plate interface between the 1971a and 1975a earthquakes is presently locked, and therefore the probable site of an imminent large earthquake, is provided by examining the state of stress in its outer rise. Several large outer rise earthquakes with $M_s \geq 5.4$ and with known focal mechanisms have occurred in this region. Locations and mechanisms for these events are shown in Figure 13. No compressional outer rise events have occurred in this vicinity in the last sixty years. All the outer rise events excepting those in 1967 and 1937 (no focal mechanism available) have followed large underthrusting earthquakes, when we expect the outer rise to be dominantly in tension. Both the 1937 and 1967 events occur in the outer rise directly in front of the large earthquake gap. Since the 1937 event is the larger of the two, it is important to determine its focal mechanism to adequately characterize stresses in the outer rise. Figure 14 shows first-motions for the 1937 event from stations listed in the ISS catalogue. Although neither nodal planes are well determined, all first motions in the center of the focal sphere are dilatational, indicating that the 1937 event is a tensional outer-rise earthquake. Although many large subduction zone events have occurred without preceding compressional outer-rise seismicity, the lack of compression building in the outer rise and the absence of $M_s \geq 7.0$ earthquakes during the last two recurrence periods (50–80 years) causes us to favor the interpretation that this region of the plate interface is weakly coupled and subducting largely aseismically.

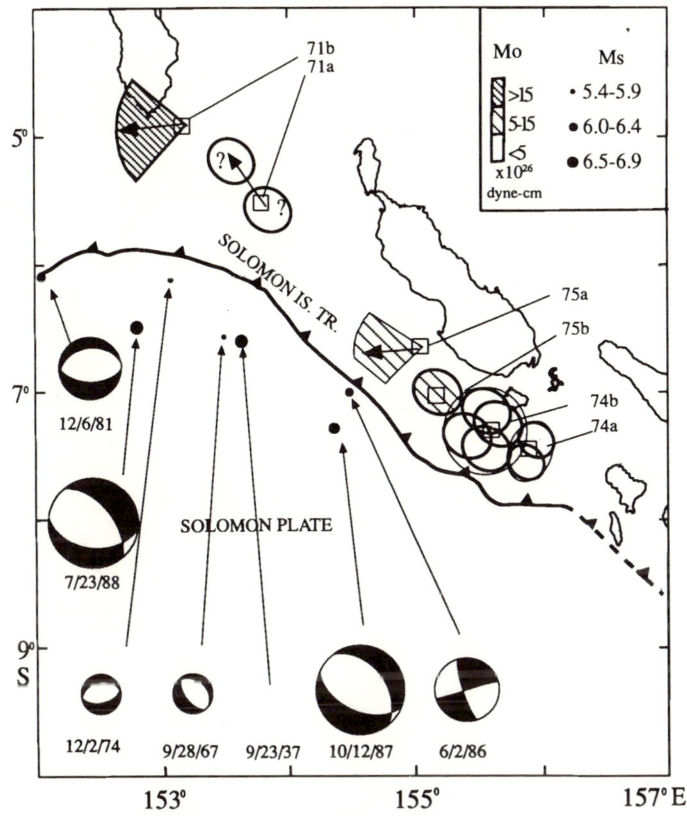

Figure 13

Locations and focal mechanism solutions for outer-rise earthquakes ($M_s \geq 5.4$) since the 1960s in the northern Solomon Islands Arc. Epicentral symbols and focal mechanisms are scaled to magnitude. The event on 9/23/1937 is $M_s = 6.9$. All the outer-rise events with tensional mechanisms followed a large underthrusting earthquake with the exception of the 1937 (first motions are shown in Figure 14) and 1967 events. These two events indicate that the outer rise in front of the large earthquake gap is in tension and that a large underthrusting earthquake is probably not imminent.

Conclusions

Detailed body wave studies of the 1974 and 1975 northern Solomon Islands doublet earthquakes indicate a large variation in their rupture process. The smaller 1974 events ($M_w = 7.3$ and 7.4) are more complicated with long source durations consisting of several subevents. The second of the 1975 doublet earthquakes ($M_w = 7.3$) is well modeled by a simple source rupturing an approximate 20 km radius about its epicenter. Only the first of the 1975 events ($M_w = 7.6$) exhibits source directivity requiring rupture of a finite fault. This event ruptured primarily updip, with at least half of its body wave moment released 50 km from the epicenter. Relative static stress drops estimated from body wave source areas and

September 23, 1937

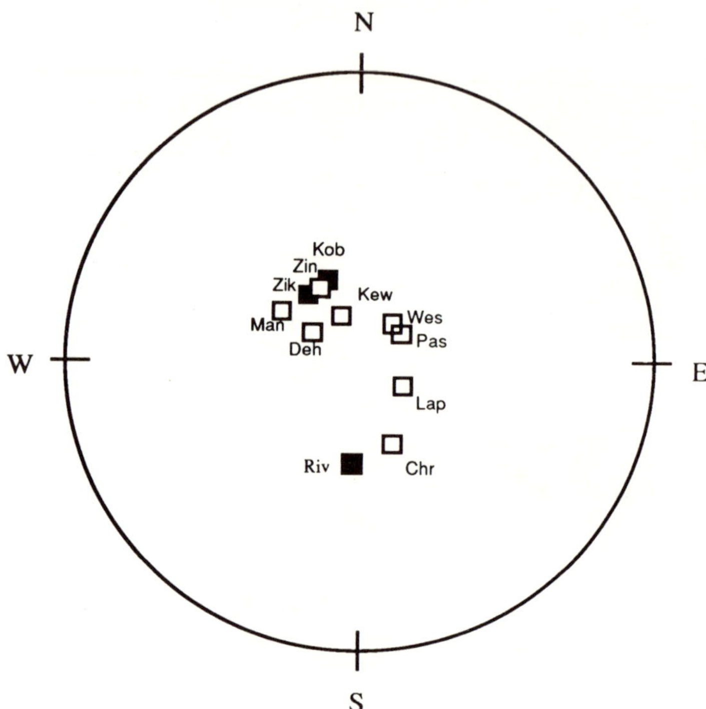

Figure 14

First motions for the 1937 outer rise earthquake taken from the ISS catalogue. Dilatational motions (open symbols) in the center of the lower hemisphere projection suggest that this earthquake had a tensional mechanism.

seismic moments associated with all six northern Solomon Islands doublet events (including the largest 1971 doublet studied by SCHWARTZ *et al.*, 1989a) allow us to compare fault properties along the plate interface. We find that the 1974 events have lower stress drops than the 1971b and 1975 earthquakes, which we interpret as rupturing weaker portions of the plate interface. The transition in mechanical coupling of the plate interface, as the 1974 source region is approached, may be related to its proximity to subduction of the topographically high Woodlark Rise or actively spreading Woodlark Ridge.

Relocation of 85 underthrusting earthquakes in the northern Solomon Islands Arc defines the seismogenic interface to be approximately 90 km wide, extending in depth from near the surface to about 65 km. The position of the mainshock asperity regions within the seismogenic zone reveals that few smaller magnitude events overlap asperity regions. The majority of small magnitude underthrusting earthquakes occupy a segment of the plate interface that has never experienced a magnitude greater than 7.0 earthquake in historic times. The relatively small recurrence interval for large underthrusting earthquakes in adjacent sections of the northern Solomon Islands Arc (50–80 years) emphasizes the anomalous nature of this segment. Tensional stresses in the outer rise just seaward of this region, determined from focal mechanisms of several recent earthquakes, suggest that this portion of the plate interface is not locked and is most likely slipping aseismically.

Acknowledgments

We thank Helen Anderson and Thorne Lay for helpful discussions during this project and Renata Dmowska, Masayuki Kikuchi and an anonymous reviewer for their thoughtful and useful comments on the manuscript. This work was supported by NSF grant EAR9018487 and by a grant from the W. M. Keck Foundation. Contribution 163 of the Charles F. Richter Seismological Laboratory and the Institute of Tectonics.

REFERENCES

ABE, K. (1981), *Magnitudes of Large Shallow Earthquakes from 1904 to 1980*, Phys. Earth and Planet. Int. *27*, 72–92.

ANDO, M. (1975), *Source Mechanisms and Tectonic Significance of Historic Earthquakes Along the Nankai Trough, Japan*, Tectonophysics *27*, 119–140.

CHRISTENSEN, D. H., and RUFF, L. J. (1983), *Outer Rise Earthquakes and Seismic Coupling*, Geophys. Res. Lett. *10*, 697–700.

CHRISTENSEN, D. H., and RUFF, L. J. (1988), *Seismic Coupling and Outer Rise Earthquakes*, J. Geophys. Res. *93*, 13,421–13,444.

CHRISTENSEN, D., BECK, S., and GALEWSKY, J. (1991), *The January 14, 1976 Kermadec Earthquake Sequence*, Trans. Am. Geophys. Union *72*, 348.

COOPER, P. A. (1985), *Seismicity, Focal Mechanisms and Morphology of Subducted Lithosphere in the Papua New Guinea–Solomon Islands Region*, Ph.D. Dissertation, Univ. of Hawaii.

DEWEY, J. W. (1972), *Seismicity and Tectonics of Western Venezuela*, Bull. Seismol. Soc. Am. *62*, 1711–1751.

DMOWSKA, R., and LOVISON, L. C. (1988), *Intermediate-term Seismic Precursors for Some Coupled Subduction Zones*, Pure and Appl. Geophys. *126*, 643–664.

DMOWSKA, R., and LOVISON, L. C. (1992), *Influence of Asperities along Subduction Interfaces on the Stressing and Seismicity of Adjacent Areas*, Tectonophysics *211*, 23–43.

DZIEWONSKI, A. M., and ANDERSON, D. L. (1983), *Travel Times and Station Corrections for P Waves at Teleseismic Distances*, J. Geophys. Res. *88*, 3295–3314.

GUTENBERG, B., and RICHTER, C. F., *Seismicity of the Earth*, 2nd ed. (Princeton University Press, New Jersey 1954) pp. 161–234.

HALUNEN, A. J., and VON HERZEN, R. P. (1973), *Heat Flow in the Western Equatorial Pacific*, J. Geophys. Res. *78*, 5195–5208.

HARTZELL, S., and IIDA, M. (1990), *Source Complexity of the 1987 Whittier Narrows, California, Earthquake from the Inversion of Strong Motion Records*, J. Geophys. Res. *95*, 12,475–12,485.

HOUSTON, H., and ENGDAHL, E. R. (1989), *A Comparison of the Spatio-temporal Distribution of Moment Release for the 1986 Andreanof Islands Earthquake with Relocated Seismicity*, Geophys. Res. Lett. *16*, 1421–1424.

JOHNSON, T., and MOLNAR, P. (1972), *Focal Mechanisms and Plate Tectonics of the Southwest Pacific*, J. Geophys. Res. *77*, 5000–5032.

KANAMORI, H., and ANDERSON, D. L. (1975), *Theoretical Basis of Some Empirical Relations in Seismology*, Bull. Seimol. Soc. Am. *65*, 1073–1095.

KANAMORI, H., and STEWART, G. S. (1976), *Mode of Strain Release along the Gibbs Fracture Zone, Mid-Atlantic Ridge*, Phys. Earth Planet. Int. *11*, 312–332.

KIKUCHI, M., and KANAMORI, H. (1982), *Inversion of Complex Body Wave*, Bull. Seismol. Soc. Am. *72*, 491–506.

KIKUCHI, M., and KANAMORI, H. (1986), *Inversion of Complex Body Waves II*, Phys. Earth Planet. Int. *43*, 205–222.

LAY, T., and KANAMORI, H. (1980), *Earthquake Doublets in the Solomon Islands*, Phys. Earth Planet. Int. *21*, 283–304.

MCCANN, W. R. (1980), *Large and Moderate-sized Earthquakes: Their Relationship to the Tectonics of Subduction*, Ph.D. Dissertation, Columbia University, New York.

MENDOZA, C., and HARTZELL, S. H. (1988), *Aftershock Patterns and Mainshock Faulting*, Bull. Seismol. Soc. Am. *78*, 1438–1449.

OPPENHEIMER, D. H., BAKUN, W. H., and LINDH, A. G. (1990), *Slip Partitioning of the Calaveras Fault, California, and Prospects for Future Earthquakes*, J. Geophys. Res. *95*, 8483–8498.

PASCAL, G. (1979), *Seismotectonics of the Papua New Guinea–Solomon Islands Region*, Tectonophysics *57*, 7–34.

RICHTER, C. F., *Elementary Seismology* (W. H. Freeman & Co., San Francisco, California 1958) 768 pp.

RIPPER, I. (1975), *Some Earthquake Focal Mechanisms in the New Guinea/Solomon Islands Region, 1963–1968*, Bureau of Mineral Resources, Australia, Report no. 178, 120 pp.

RUFF, L., and KANAMORI, H. (1980), *Seismicity and the Subduction Process*, Phys. Earth Planet. Int. *23*, 240–252.

SCHWARTZ, S. Y., LAY, T., and RUFF, L. J. (1989a), *Source Process of the Great 1971 Solomon Islands Doublet*, Phys. Earth Planet. Int. *56*, 294–310.

SCHWARTZ, S. Y., DEWEY, J. W., and LAY, T. (1989b), *Influence of Fault Plane Heterogeneity on the Seismic Behavior in the Southern Kurile Islands Arc*, J. Geophys. Res. *94*, 5637–5649.

TAYLOR, B., and EXON, N. F., *An investigation of ridge subduction in the Woodlark–Solomons region: introduction and overview. In Marine Geology, Geophysics, and Geochemistry of the Woodlark Basin–Solomon Islands* (Taylor, B., and Exon, N. F. eds.) (Circum-Pacific Council for Energy and Mineral Resources, Houston, Texas 1987) pp. 1–24.

VIDALE, J., and KANAMORI, H. (1983), *The October (1980) Earthquake Sequence Near the New Hebrides*, Geophys. Res. Lett. *10*, 1137–1140.

WEISSEL, J. K., TAYLOR, B., and KARNER, G. D. (1982), *The Opening of the Woodlark Basin, Subduction of the Woodlark Spreading System, and the Evolution of Northern Melanesia since Mid-Pliocene Time*, Tectonophysics *87*, 253–277.

WESNOUSKY, G. S., ASTIZ, L., and KANAMORI, H. (1986), *Earthquake Multiples in the Southeastern Solomon Islands*, Phys. Earth Planet. Int. *44*, 304–318.

ZHANG, Z., and SCHWARTZ, S. Y. (1992), *Depth Distribution of Moment Release in Underthrusting Earthquakes at Subduction Zones*, J. Geophys. Res. *97*, 537–544.

(Received March 20, 1992, revised/accepted January 11, 1993)

Reprints from PAGEOPH

Birkhäuser

Birkhäuser Verlag AG
Basel Boston Berlin